42

Advances in Polymer Science
Fortschritte der Hochpolymeren-Forschung

Editors: H.-J. Cantow, Freiburg i. Br. · G. Dall'Asta, Colleferro · K. Dušek, Prague · J. D. Ferry, Madison · H. Fujita, Osaka · M. Gordon, Colchester
J. P. Kennedy, Akron · W. Kern, Mainz · S. Okamura, Kyoto
C. G. Overberger, Ann Arbor · T. Saegusa, Kyoto · G. V. Schulz, Mainz
W. P. Slichter, Murray Hill · J. K. Stille, Fort Collins

New Polymerization Reactions

With Contributions by
I. P. Breusova, H. K. Hall, M. Hasegawa
D. S. Johnston, B. P. Morin, Z. A. Rogovin
Y. Yokoyama

With 26 Figures

Springer-Verlag
Berlin Heidelberg New York 1982

Editors

Prof. Hans-Joachim Cantow, Institut für Makromolekulare Chemie der Universität, Stefan-Meier-Str. 31, 7800 Freiburg i. Br., BRD

Prof. Gino Dall'Asta, SNIA VISCOSA – Centro Studi Chimico, Colleferro (Roma), Italia

Prof. Karel Dušek, Institute of Macromolecular Chemistry, Czechoslovak Academy of Sciences, 162 06 Prague 616, ČSSR

Prof. John D. Ferry, Department of Chemistry, The University of Wisconsin, Madison, Wisconsin 53706, U.S.A.

Prof. Hiroshi Fujita, Department of Polymer Science, Osaka University, Toyonaka, Osaka, Japan

Prof. Manfred Gordon, Department of Chemistry, University of Essex, Wivenhoe Park, Colchester C04 3 SQ, England

Prof. Joseph P. Kennedy, Institute of Polymer Science, The University of Akron, Akron, Ohio 44325, U.S.A.

Prof. Werner Kern, Institut für Organische Chemie der Universität, 6500 Mainz, BRD

Prof. Seizo Okamura, No. 24, Minami-Goshomachi, Okazaki, Sakyo-Ku, Kyoto 606, Japan

Prof. Charles G. Overberger, Department of Chemistry, The University of Michigan, Ann Arbor, Michigan 48 104, U.S.A.

Prof. Takeo Saegusa, Department of Synthetic Chemistry, Faculty of Engineering, Kyoto University, Kyoto, Japan

Prof. Günter Victor Schulz, Institut für Physikalische Chemie der Universität, 6500 Mainz, BRD

Dr. William P. Slichter, Chemical Physics Research Department, Bell Telephone Laboratories, Murray Hill, New Jersey 07 971, U.S.A.

Prof. John K. Stille, Department of Chemistry, Colorado State University, Fort Collins, Colorado 805 23, U.S.A.

ISBN-3-540-10958-7 Springer-Verlag Berlin Heidelberg New York
ISBN-0-387-10958-7 Springer-Verlag New York Heidelberg Berlin

Library of Congress Catalog Card Number 61-642

This work is subject to copyright. All rights are reserved, whether the whole or part of the material is concerned, specifically those of translation, reprinting, re-use of illustrations, broadcasting, reproduction by photocopying machine or similar means, and storage in data banks. Under § 54 of the German Copyright Law where copies are made for other than private use, a fee is payable to the publisher, the amount to "Verwertungsgesellschaft Wort", Munich.

© Springer-Verlag Berlin Heidelberg 1982
Printed in Germany

The use of general descriptive names, trademarks, etc. in this publication, even if the former are not especially identified, is not to be taken as a sign that such names, as understood by the Trade Marks and Merchandise Marks Act, may accordingly be used freely by anyone.
Typesetting and printing: Schwetzinger Verlagsdruckerei. Bookbinding: Brühlsche Universitätsdruckerei, Gießen.
2152/3140 – 543210

Table of Contents

Four-Center Photopolymerization in the Crystalline State
   M. Hasegawa 1

Macrozwitterion Polymerization
   D. S. Johnston 51

Ring-Opening Polymerization of Atom-Bridged and Bond-Bridged Bicyclic Ethers, Acetals and Orthoesters
   Y. Yokoyama and H. K. Hall 107

Structural and Chemical Modifications of Cellulose by Graft Copolymerization
   B. P. Morin, I. P. Breusova and Z. A. Rogovin 139

Author Index Volumes 1–42 167

Four-Center Photopolymerization in the Crystalline State

Masaki Hasegawa

Faculty of Engineering, The University of Tokyo, Tokyo, Japan

A typical reaction of four-center photopolymerization is that of 2,5-distyrylpyrazine (DSP), which was reported by the author in 1967.

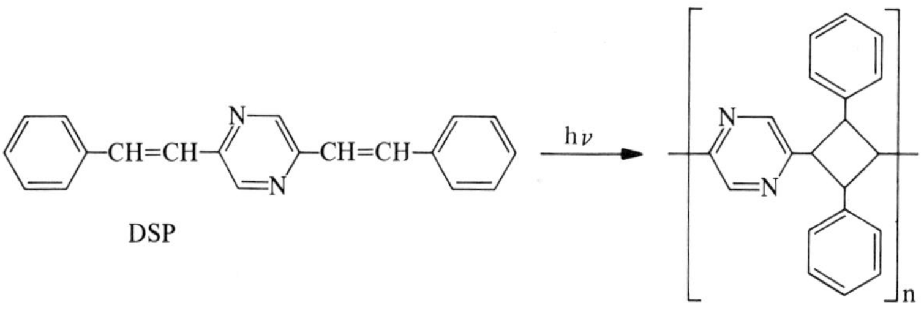

Since then, a large number of new polymer crystals have been prepared from conjugated diolefin crystals. All the polymers studied so far, are highly crystalline and three dimensionally oriented with the 1,3-trans cyclobutane and 1,4-phenylene groups alternating in the main chain.

In 1978 Lahav et al. (Weizmann Institute, Israel) succeeded in an absolute asymmetric syntheses of chiral oligomeric crystals by the four-center polymerization of achiral monomer.

X-ray crystallographic studies during the course of the reaction have demonstrated that the reaction is a typical topochemical process involving a direct rearrangement of the monomer crystal to the polymer crystal having an extended rigid rod-like structure. By x-ray analysis and DSC on the thermal depolymerization of the polymer crystal, a reversible topochemical process has been demonstrated for monomer and polymer crystals.

Considering these characteristics from an overall point of view, it is assumed that in the four-center photopolymerization, both polymerization process and polymer properties can provide suitable patterns for naturally occurring polymers.

I.	Introduction	3
II.	**Preparation and Chemical Structure of Polymers**	5
	a. Preparation of Polymers	5
	b. Chemical Structure of Polymers	13

III.	**Polymerization Mechanism**	17
	a. Polymerization Kinetics	17
	b. Effect of Wavelength of Exciting Light	20
	c. Quantum Yield	23
	d. Matrix Effect. Solution Photopolymerization and Photodepolymerization	24
IV.	**Crystallography**	27
	a. Molecular Alignment and Relative Orientation of Monomer and Polymer Crystals	28
	b. Polymerization Mechanism Based on Topotaxy	32
V.	**Changes of Morphology and Thermodiagram in the Course of Polymerization**	34
	a. Morphological Changes in the Polymerization	34
	b. Changes in the Thermodiagram during Polymerization	36
VI.	**Polymer Properties**	40
	a. Chemical Properties	40
	b. Physical Properties	44
VII.	**Characteristic Features of Four-Center Photopolymerization**	45
VIII.	**References**	48

I. Introduction

Organic photoreactions in the crystalline state have been studied widely and date back to the end of the last century.

For example, the photoreaction of cinnamic acid crystals was described by Liebermann in 1889[1]. Since then, a great variety of crystalline state reactions have been reported, e. g. dimerization, cis-trans isomerization, substituent migration, and formation of "cage" molecules.

The crystal structures of the starting compounds play an important role in the reaction. The determination of the steric configuration of the products as well as the crystal structures of reactant and product and the crystal change during the course of reaction provide important information for the understanding of crystalline state reactions. Crystalline state photoreactions of cinnamic acid derivatives and some other olefinic compounds were elucidated in 1964 by Schmidt and co-workers[2] who correlated the crystal structure of the reactant with the photoreactivity and steric configuration of the products. According to the results obtained, the crystal structures of trans-cinnamic acid derivatives are divided into three types, i.e. α-, β-, and γ-type crystal packings where the intermolecular distances between olefinic double bonds are 3.6–4.1, 3.9–4.1, and 4.7–5.1 Å, respectively. A typical example is the photoirradiation of trans-cinnamic acid where the olefinic double bonds react in the crystal to form cyclobutane, resulting in the formation of α-truxillic acid from α-type crystals and in β-truxinic acid from β-type crystals whereas the γ-type crystals of cinnamic acid derivatives are stable to light since the olefinic double bonds are placed at distances too long to undergo reactions with other double bonds in the crystal. A correlation between molecular alignment in the reactant crystal and steric configuration of the product is explained quite reasonably in terms of "topochemical control".

Solid-state polymerization has been investigated mostly for chain reactions initiated by γ-rays, light, catalysts or heat, for more than twenty years. On the other hand, the stepwise crystalline-state photopolymerization had not been reported before 1967, though the crystalline-state organic photoreactions had been a familiar phenomenon for many years.

In 1958 Koelsch and Gumprecht reported that brilliant yellow crystals of 2,5-distyrylpyrazine became white on exposure to UV irradiation, turning into an insoluble polymeric substance with a melting point of 331–333 °C[3]. A few years later, the present group of authors independently made the same observation with 2,5-distyrylpyrazine on exposure to sunlight during storage in the course of a preparatory study of pyrazine-2,5-dicarboxylic acid from 2,5-distyrylpyrazine, and reported it in 1965[4]. This phenomenon was further investigated, and it was concluded that a linear high molecular weight polymer with cyclobutane units in its main chain had been produced from 2,5-distyrylpyrazine crystals by the action of UV light or sunlight. This new type of reaction, which is called four-center type photopolymerization, was reported in Polymer Letters in 1967[5].

Also in 1967, Holm and Zienty announced in patent that crystalline linear polymers were obtained by photoirradiation of a number of crystalline dibenzylidenebenzene diacetonitrile derivatives[6]. In 1972, they published the results on the photopolymerization of dibenzylidenebenzene diacetonitrile in J. Polymer Science[7].

Four-center photopolymerization in the crystalline state is a general term for the reactions in which conjugated diolefin crystals are photochemically converted in a stepwise manner into crystals of linear polymers containing cyclobutane rings in their main chain.

$$\text{2,5-Distyrylpyrazine} \xrightarrow{h\nu} \text{Poly-2,5-distyrylpyrazine}$$

2,5-Distyrylpyrazine   Poly-2,5-distyrylpyrazine

Since investigations on diolefinic compounds such as 2,5-distyrylpyrazine or dibenzylidenebenzene diacetonitrile to be used as starting monomers for the synthesis of linear high molecular weight polymers had previously not been reported, these findings were of great significance, and research work in this field entered a new stage of development. The authors have explored the photopolymerizability of various types of diolefinic compounds in the crystalline state by irradiation with UV and visible light and synthesized a large number of crystalline high molecular weight polymers of linear structure.

For the past fourteen years a wide variety of studies have been carried out on the polymerization kinetics of diolefinic monomers, the correlations between the crystal structure of the monomer and that of the polymer, the morphology of the reacting crystals, the crystal lattice effect on the polymer properties, and the thermal depolymerization in the crystalline state. On the basis of these studies, it has been concluded that the four-center photopolymerization in the crystalline state is a new type of topochemical reactions involving direct rearrangement of monomer crystals to polymer crystals under the strict control of the reacting crystal lattice. In addition, the reaction only proceeds with favorably aligned molecules in the monomer crystal, resulting in the formation of extremely rigid rod-shaped polymer chain[8].

In 1978, Lahav et al. succeeded in performing an absolute asymmetric synthesis of chiral dimers and oligomers through crystallization of an achiral monomer in a chiral crystal, followed by the four-center type photopolymerization in the crystalline state[9].

In a recent paper by the present author, based on all the results obtained so far, the characteristic features of the four-center photopolymerization of crystalline diolefinic monomers and of the properties of the resulting polymers have been discussed with particular emphasis on the dependence of thermal properties on morphology[10]. Furthermore, thermal stability which depends on molecular weight was attributed to the rigid rod-shaped polymer chain structure. In summary, the results of the studies on the thermal behavior of the obtained crystalline polymers and the polymerization process have been recognized to be significant for the design of models of naturally occurring polymers.

Only a few review articles concerning this type of topochemical polymerization have been published so far[8, 11] from the viewpoint of a specific field, though extensive work has been done in recent years. This article is the first review of a four-center photopolymerization and related problems which covers all the results obtained from a variety of fields since 1967.

II. Preparation and Chemical Structure of Polymers

2,5-Distyrylpyrazine (DSP), which was first prepared by Franke in 1905[12], is the first crystalline monomer found capable of undergoing a four-center photopolymerization. Bright yellow crystals of DSP with a melting point of 236 °C, when exposed to sunlight, e.g. for a day in a rotating glass flask, turn into a powdery white crystalline high molecular weight linear polymer. Poly-DSP has an extremely high degree of crystallinity and is observed as brilliant crystals under a polarizing microscope. An amorphous transparent film is prepared from poly-DSP by casting from solution[13]. The same type of crystalline-state photopolymerization was discovered for a series of α,α'-bis(4-acetoxy-3-methoxy-benzylidene)-p-benzenediacetonitrile (AMBBA) by Holm and Zienty, which was published as a patent in 1967[6].

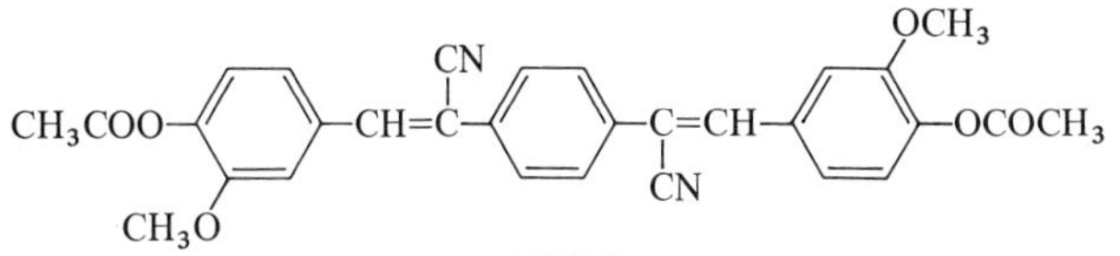

AMBBA

These two discoveries of the new type of crystalline state photopolymerization prompted polymer chemists to study the generality of this type of polymerization. In consequence, a large number of diolefinic compounds have been found to photopolymerize to linear high molecular weight polymers by a four-center photopolymerization in the crystalline state. All the diolefinic monomer crystals investigated so far have been found to undergo no detectable polymerization upon prolonged irradiation with γ- or X-rays.

a. Preparation of Polymers

Subsequent to the discovery of the polymerization of DSP crystals, another monomer, 1,4-bis(3-pyridyl-2-vinyl)benzene (P2VB), which belongs to the DSP series (Ar'–CH=CH–Ar–CH=CH–Ar'), was found to photopolymerize to the same type of polymer in a similar manner. However, P2VB polymerizes much more slowly than DSP[13]. Polymerization of the former can be regarded as an extension of the dimerization of stilbazol in the crystalline state[14].

Then, the photopolymerizability of p-phenylenediacrylic acid (PDA) and its derivatives, which possess two cinamic acid units in a single molecule, was investigated for comparison with the crystalline-state photodimerization of cinnamic acid[15].

Similar series of olefins such as α,α'-dicyano-p-phenylenediacrylic acid (CPAS)[16], p-phenylenedibutadienoic acid (PDBA), p-phenylene-bis-(α-cyanobutadienoic acid) (PDCBA)[17], and their derivatives have been prepared and subjected to photopolymerization in the crystalline state.

IR and NMR spectroscopy revealed that most of these compounds possess a trans-,trans-conformation and this was confirmed also for several photopolymerizable crystalline monomers by X-ray crystallographic analysis (see Sect. IV. a).

Since all the monomers absorb light of wavelengths from the ultraviolet to the visible region, polymerization proceeds by irradiation with a xenon or a high-pressure mercury lamp, and sometimes even rapidly on exposure to sunlight.

$$Ar'-CH=CH-Ar-CH=CH-A'r$$

DSP series

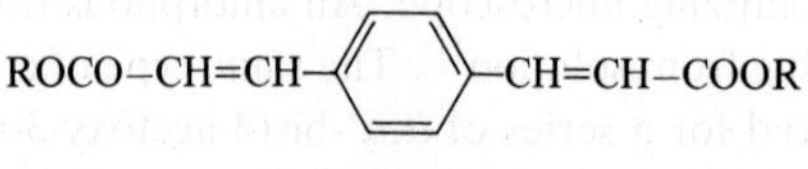

$$ROCO-CH=CH-\underset{}{\bigcirc}-CH=CH-COOR$$

PDA series

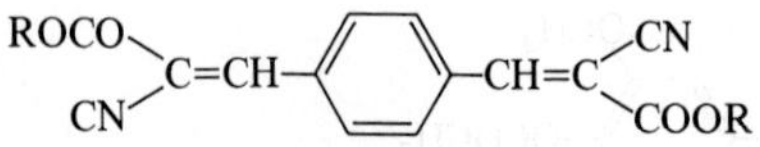

$$\underset{CN}{ROCO}\!\!>\!\!C=CH-\underset{}{\bigcirc}-CH=C\!\!<\!\!\underset{COOR}{CN}$$

CPA series

$$ROCO-CH=CH-CH=CH-\underset{}{\bigcirc}-CH=CH-CH=CH-COOH$$

PDBA series

$$\underset{CN}{ROCO}\!\!>\!\!C=CH-CH=CH-\underset{}{\bigcirc}-CH=HC-CH=C\!\!<\!\!\underset{COOR}{CN}$$

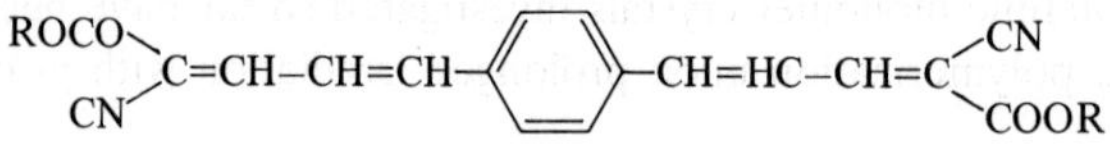

PDCBA series

In order to check photopolymerizability, each monomer is dispersed as fine crystals in a suitable inert dispersant such as water or water-ethanol in a quartz flask and irradiated with an appropriate light source for a period of several minutes to a few days at room temperature or lower with stirring. Several monomer crystals are orange-yellow and fade to opaque white during the reaction. Polymerization proceeds smoothly at a temperature considerably lower than the melting point of the monomer.

The potassium bromide-pellet method provides the most convenient way to detect a slight polymerizability at room temperature using the following operation: a few milligrams of the monomer are embedded in a KBr pellet, and the change upon irradiation is followed by IR spectroscopy. All the polymers thus obtained, their properties and the polymerization conditions are listed in Table 1.

It should be noted that all the polymer crystals are derived from monomers that have a rigid linear structure with conjugated double bonds separated by the 1,4-position (2,6-position in naphthalene of NBA-DCP) of an aromatic ring. Polymerization proceeds with a high selectivity at a suitable temperature, and the monomer molecules grow into a linear polymer at high conversion in most of the cases shown in Table 1; the polymerization behavior is greatly affected by the reaction temperature especially near the melting or crystal transition point (see Sect. III. a.)[19].

Table 1. Polymerization conditions and polymer properties[a]

	Reactivity	Reaction temp. (°C)	Polymer yield (%)	Decomp. point[b] (°C)	Reduced viscosity (dl/g)	Example of solvents for as-polym. polymer crystals	Ref.
General formula of polymers: $\left[-Ar-\underset{R'R}{\overset{RR'}{\Box}} - \right]_n$	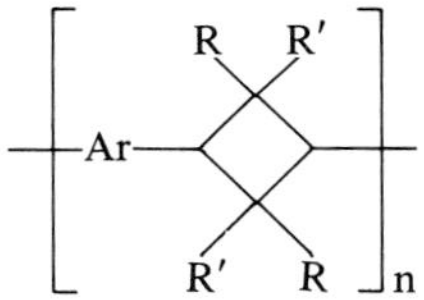						
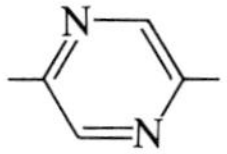 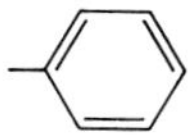 Ar: pyrazine R: phenyl, R': –H	very high	room	quantitative[c]	339–343	1.0–10	conc · H_2SO_4 CF_3COOH $CHCl_2COOH$ (m-cresol, o-chlorophenol)[d]	13)
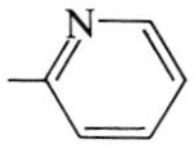 Ar:  phenylene R: pyridyl, R': –H	low	room	quantitative	340	0.3–2	(m-cresol, o-chlorophenol)[d]	13)
Ar:  phenylene R: –COOH, R': –H	low	room	quantitative	290	0.12	conc · H_2SO_4	15)

Table 1 (continued)

Ar / R / R'	Reactivity	Reaction temp. (°C)	Polymer yield (%)	Decomp. point[b] (°C)	Reduced viscosity (dl/g)	Example of solvents for as-polym. polymer crystals	Ref.
Ar:  ⌬ (p-phenylene) R: –COOCH$_3$, R': –H	high	room	quantitative	415	>10	conc · H$_2$SO$_4$	15)
Ar:  ⌬ R: –COOC$_2$H$_5$ R': –H	high low	–25 room	quantitative high	347 –	1.4 0.16	conc · H$_2$SO$_4$	15)
Ar:  ⌬ R: –COO–n–C$_3$H$_7$ R': –H	low	0–5	42	360	0.26	conc · H$_2$SO$_4$	15)
Ar:  ⌬ R: –COO–i–C$_3$H$_7$ R': –H	low	room	65	320	0.45	conc · H$_2$SO$_4$	15)
Ar:  ⌬ R: COOC$_6$H$_5$ R': –H	medium	room	85	420	0.90	aq. alkali (accompanied by hydrolysis)	15)

Ar, R, R'					
Ar:  ⌬ R: –CONH$_2$ R': –H	medium	room	quantitative	405	1.50	conc · H$_2$SO$_4$	15)
Ar:  ⌬ R: –COOCH$_3$ R': –CN	high	room	quantitative	290	0.3	conc · H$_2$SO$_4$	16)
Ar:  ⌬ R: –COOC$_2$H$_5$ R': –CN	low	room	66	340	2.6	conc · H$_2$SO$_4$	16)
Ar:  ⌬ R: –COO–n–C$_3$H$_7$ R': –CN	very high	room	quantitative	335	3.0	conc · H$_2$SO$_4$	16)
Ar:  ⌬ R: –COO–i–C$_3$H$_7$ R': –CN	low	room	quantitative	320	1.8	conc · H$_2$SO$_4$	16)

Table 1 (continued)

	Reactivity	Reaction temp. (°C)	Polymer yield (%)	Decomp. point[b] (°C)	Reduced viscosity (dl/g)	Example of solvents for as-polym. polymer crystals	Ref.
Ar:  –C$_6$H$_4$–CH$_3$ R: –COO–n–C$_4$H$_9$ R': –CN	low	room	quantitative	330	0.6	conc · H$_2$SO$_4$	16)
Ar:  –C$_6$H$_4$–CH$_3$ R: –COO–n–C$_8$H$_{17}$ R': –CN	low	room	quantitative	–	–	–	16)
Ar:  –C$_6$H$_4$–CH$_3$ R: –CH=C(CN)(COOC$_2$H$_5$) R': –H	high	room	quantitative	245	1.3 [η]	conc · H$_2$SO$_4$ CF$_3$COOH	17)
 Ar: –C$_6$H$_4$–CH$_3$ R: –CH=C(CN)(CONH$_2$) R': –H	low	room	–	–	–	–	17)

Ar: [2,6-disubstituted naphthalene]

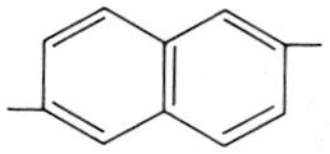

R: –COO–[3,5-dichlorophenyl]

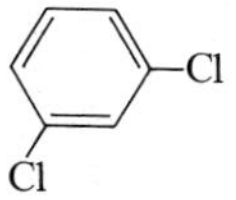

R': –H

— — — — — 18)

General formula of polymers:

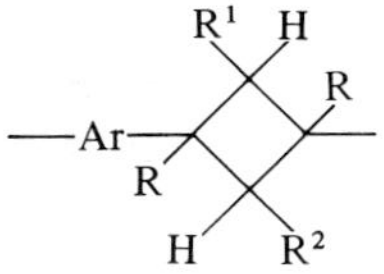

$$-\text{Ar}-\underset{\underset{H}{\overset{R}{|}}}{\overset{\overset{R^1}{|}}{C}}-\underset{\underset{R^2}{\overset{H}{|}}}{\overset{\overset{R}{|}}{C}}-$$

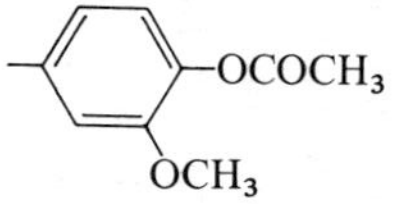

high   room   quantitative[c]   300   0.44 [η]   —

Ar: [p-substituted phenyl]

R: –CN

R¹, R²: –OCOCH₃, –OCH₃

conc · H_2SO_4 7)[e]

[a] Results are not always given for the optimum conditions of polymerization but for one of the typical experimental runs
[b] Measured by capillary method. Decomposition point is not strictly definable because almost all as-polymerized polymer crystals thermally depolymerize in the crystalline state, and the starting point of depolymerization is dependent on their molecular weight
[c] Photostable crystal is obtained by recrystallization under a specified condition
[d] Solvents in parenthesis are the solvents used only for amorphous polymer
[e] In the patent literature[6], a large number of crystals of the AMBBA series are reported to be photopolymerizable

DSP[13] and p-CPA nPr[16] have apparently the highest photoreactivity among the monomers listed in Table 1, while P2VB[13] and PDA amide[15] polymerize rather slowly (Sect. III. c.). In a few cases (DSP[20] and AMBBA[7]), the same compound with different crystal forms shows different photochemical behaviors, i.e. one form is photopolymerizable and the other one is photostable. In the course of p-CPA Me[16] polymerization, cis, trans-isomerization in the reacting crystals is observed in the NMR spectrum. This could explain the depression both of the chain length and crystallinity of the resulting polymer.

Molecules with a certain flexibility such as poly(methylene-bis-cinnamate) or poly-(cinnamic anhydride) do not undergo polymerization to linear high polymers in the crystalline state. Polymerization of glycol bis-cinnamates has been reported[21], although the authors could not confirm the formation of a polymer having cyclobutane rings. Interesting exceptional cases are the polymerizations of methyl m-phenyleneacrylate (m-PDA Me)[22] and pentaerythritol tetracinnamate[23] crystals. In these reactions, conversion based on the disappearance of the olefinic bonds is very high, but only amorphous products are formed. During the polymerization the apparent crystal form is unaltered in the case of pentaerythritol tetra-cinnamate whereas cracks or disintegration into fibers are observed with all other photopolymerizable monomer crystals.

Lahav et al. have obtained some oligomeric substances from ethyl-2-cyano-3 (p-sec.-butyl-3'-E-propenoate)-phenyl-E-propenoate-1 which has two different types of photodimerizable olefinic double bonds within the same molecule[9].

$$\text{Sec.BuOCOCH=CH}-\text{C}_6\text{H}_4-\text{CH=C}\begin{array}{c}\text{CN}\\\text{COOEt}\end{array}$$

Although polymerization of the monomer does not proceed to high conversion, this process is most significant because it is the first example of an asymmetric synthesis starting from racemic monomer crystals.

According to X-ray powder pattern analysis, the degree of crystallinity in the majority of the polymers obtained is as high as that of the original monomer. The X-ray diffraction pattern of poly(p-PDA amide) is shown in Fig. 1.

Most crystalline polymers do not exhibit a crystalline melting point but a decomposition point where depolymerization starts (ca. 340 °C for poly-DSP and ca. 420 °C for poly-PDA phenyl ester by DSC). The decomposition point, however, is not strictly definable, because it is greatly affected by the morphology and the molecular weight of the polymer (see Sect. VI. a.).

In solution, these polymers depolymerize both photochemically and thermally to the monomers in very high yield, or only to the oligomers in some exceptional cases (see Sect. VI. a.).

Preparative studies on four-center photopolymerization have demonstrated empirically that the crystalline state photodimerization of olefins such as stilbazol[14], cinnamic[1], α-cyanocinnamic acid[24], or 4-phenylbutadienoic acid[25] derivatives has been extended to the crystalline-state photopolymerization of rigid linear monomers having two dimerizable olefinic units in a single molecule, as is common in stepwise polymerization in liquid phase where the monofunctional reaction is generally extendable to the corresponding bifunctional reactions.

As illustrated for DSP[26] and AMBBA[7] polymorphs, nonpolymerizable monomers in one modification sometimes become polymerizable in the other crystal modifications. No

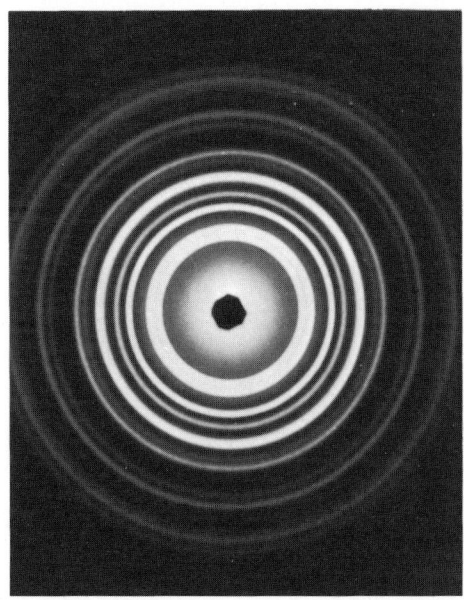

Fig. 1. X-ray diffraction photograph of poly(p-PDA Amide). (source: Ref. 15)

example of photopolymerizable crystals has been found so far as an extention of the well-known dimerization reaction of chalcone[27] to a bifunctional molecule e.g. 1,4-bis(2-benzoylvinyl)benzene. For a further discussion of this empirical rule on the extension of dimerization on the basis of crystallographic results (see Sect. IV. a.).

b. Chemical Structure of Polymers

The chemical structure of polymers has been studied by IR and NMR spectroscopy and chemical analysis. The compositions of all the photoproducts obtained are the same as

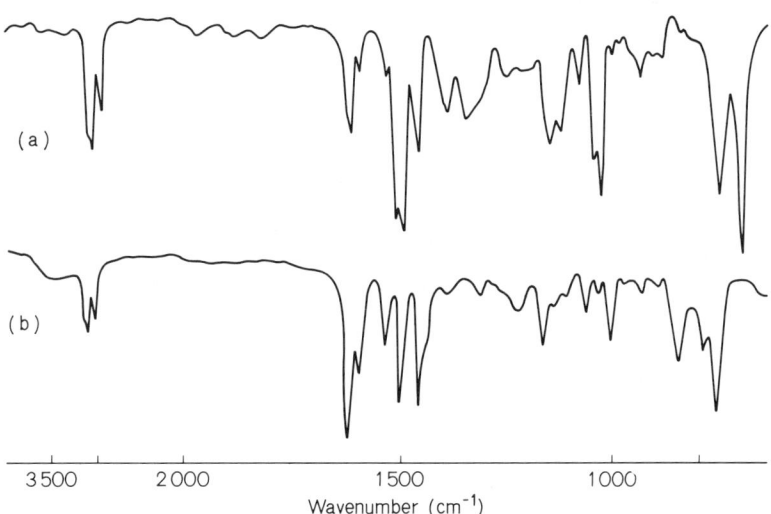

Fig. 2a, b. IR spectra of (a) poly-DSP and (b) poly-P2VB. (source: Ref. 13)

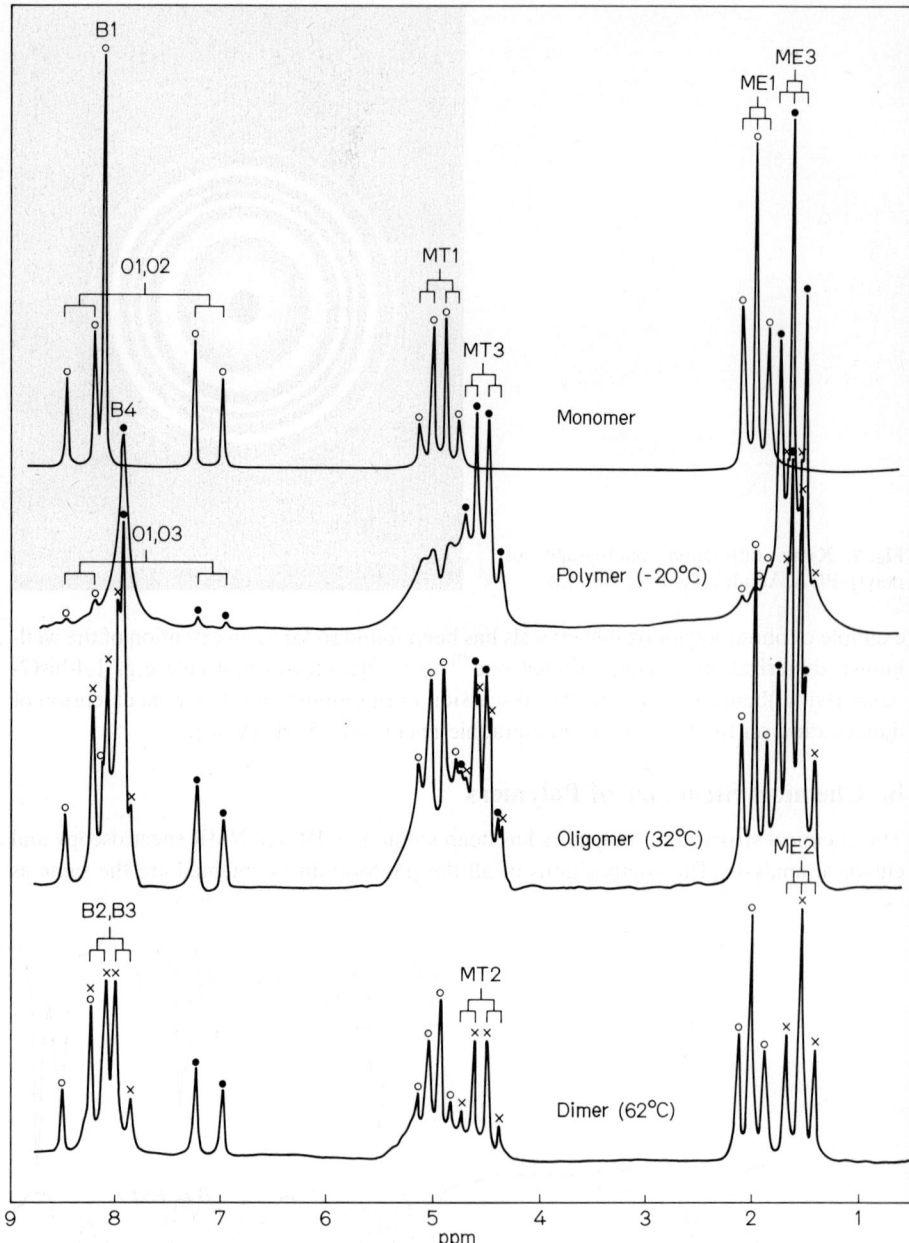

Fig. 3. NMR spectra of p-PDA Et, its oligomer, dimer and polymer, and peak assignment to their molecular structures. Figures in parentheses indicate the reaction temperature. (source: Ref. 29)

Fig. 3 (continued)

those of the monomers within experimental error. The product solution has a high viscosity and yields a clear flexible tough film when cast on a glass plate. From these results, the formation of a linear high polymer by cycloaddition of double bonds is conclusive.

The IR spectra of poly-DSP and poly-P2VB are shown in Fig. 2.

Poly-DSP contains neither peaks at 1630 cm^{-1} (C=C aliphatic) nor at 976 cm^{-1}, (C=C trans HC=HC) which are intense in the monomer. The corresponding peaks in P2VB, 1640 and 970 cm^{-1}, respectively, are also absent in poly-P2VB. The lack of olefinic double bonds is confirmed chemically by inertness of these polymers to bromine. A weak peak at 930 cm^{-1} in both polymers is attributed to cyclobutane.

The NMR spectra of poly-DSP and poly-P2VB show a broad band at τ 4.9–5.0 (4H) which is characteristic of protons bonded to a 1,2,3,4-tetraarylated cyclobutane ring, and quantitative considerations are consistent with this assignment.

Other peaks at τ 2.9, 2.8 and 1.3 for poly-DSP are assigned to 3,3'-protons (4H), 2,2'- and 4,4'-protons (6H) of benzene, and to protons (2H) at the 3- and 6-positions of pyrazine. The peaks at τ 2.8, 2.2 and 1.5–1.6 of poly-P2VB correspond to benzene protons (4H), protons (4H) at the 3,3'- and 5,5'-positions of pyridine and protons (4H) at the 4,4'- and 6,6'-positions of pyridine. From the IR and NMR spectroscopic results poly-DSP and poly-P2VB are concluded to be linear polymers with alternating recurring cyclobutane and aromatic rings in the main chain.

The absolute configuration of the polymer as well as head-to-tail and head-to-head structures have not been definitely determined by NMR spectroscopy because of the low resolution for polymers. However, the high symmetry of the NMR spectra of most polymers suggests that these polymers exhibit a single steric configuration. An exception is a small amount of cyclobutane presumably formed from the cis-form of an olefinic double bond which is seen in poly-p-CPA Me. Of the many polymer crystals as-polymerized, a 1,3-trans configuration of cyclobutane has been demonstrated by X-ray crystallographic analysis (see Sect. IV. a.).

NMR spectroscopic studies have been performed in detail on p-PDA Et, its dimer, oligomer and high polymer[19]. The dimer and the oligomer of p-PDA Et are prepared by photo-irradiation with a selected wavelength or at a definite temperature (see Sect. III. a.). NMR spectra of the monomer, the dimer, the oligomer and high polymer of p-PDA Et are shown in Fig. 3.

The three types of singlet or doublet aromatic protons correspond to aromatic protons in the monomer, the terminal unit and the repeating polymer unit, respectively. The two types of trans-olefinic protons are assigned respectively to those in the monomer and the terminal unit. The protons of the ethyl ester moiety are distinguished as three different types; those in the monomer, in the branched group attached to cyclobutane next to the terminal unit, and in the polymer chain.

An IR spectral study on the same samples provides similar structural information. Carbonyl absorption peaks appear at 1726 cm^{-1} in the polymer chain. Olefinic double bonds are distinguished between the monomer and the terminal unit by the stretching vibration peak which appears at 1633 and 1639 cm^{-1}, respectively.

For other polymers, similar IR and NMR spectroscopic results have been obtained, and a linear polymer with a cyclobutane structure is concluded as well.

III. Polymerization Mechanism

The polymerization mechanism has been studied with respect to the type of reaction and to other factors such as reaction temperature, wavelength of exciting light, quantum yield or the effect of crystal matrix.

The polymerization proceeds in a stepwise manner by cyclobutane formation between the excited olefin and the olefin in the ground state. The concentration term is not definable in the usual sense, and the reaction proceeds by favorably aligned molecules in the monomer crystal. At elevated temperatures, the apparent reaction rate is accelerated in the range below the melting point of the reacting crystals at the early stage whereas the molecular weight of the final product is diminished. The stepwise polymerization process may be separated into two processes, oligomerization and subsequent polymerization, by controlling the wavelength of the irradiating light. Quantum yields of polymerization are extremely high in some cases (1.2–1.6 of the quantum yield).

In solution the photooligomerization of the monomer and photodepolymerization of the oligomer are controlled by the wavelength of the exciting ligth.

By comparing the photochemical behavior of conjugated diolefinic monomers in the crystalline state and in solution, a crystal matrix effect on four-center type photopolymerization has been revealed. It has been concluded that high molecular weight linear polymers are produced photochemically from these monomers only by way of a crystal-lattice controlled mechanism.

a. Polymerization Kinetics

Since the reaction is affected by the wavelength of the ligth employed and monomer crystal features such as crystal size or purity, the same apparatus and monomer crystals with the same history are employed for any series of experiments involving mechanistic study.

The monomer crystals are dispersed in a definite amount of dispersant and stirred with a magnetic stirrer at constant speed. Dispersant is required to provide homogeneous irradiation of the crystal surfaces.

Kinetic plots have been obtained for the polymerization of several monomers. Time vs. conversion and conversion vs. reduced viscosity curves of p-PDA Et at various temperatures are shown in Figs. 4 and 5.

In Fig. 4, it is seen that conversion increases rapidly at the initial stage of polymerization and then gradually with irradiation time. Conversion in Figs. 4 and 5 is the amount of polymer in weight separated by means of solubility differences and is appreciably lower than the monomer consumption measured spectroscopically. All the conversions mentioned hereafter in this chapter are weight conversions unless otherwise stated. It has been confirmed in the experiments of DSP[20] that the rate of polymerization increases with the light intensity whereas the reduced viscosity at a given conversion is nearly independent of the ligth intensity.

The reduced viscosity gradually increases with increasing conversion and rises sharply above 80% conversion. The viscosity also increases with the reaction time and continues to rise on irradiation even after the conversion is completed.

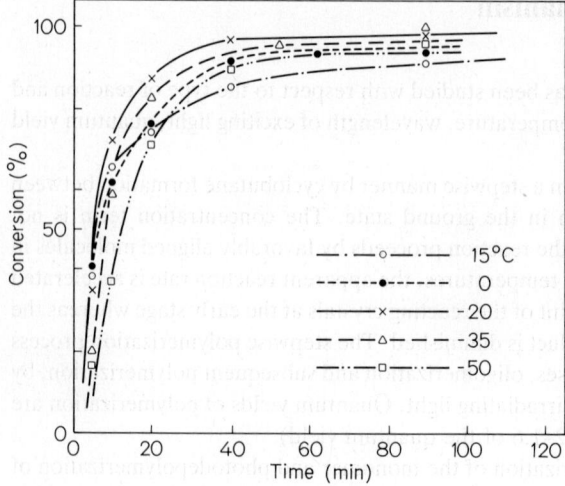

Fig. 4. Plot of time vs. conversion for polymerization of p-PDA Et at various temperatures. (source: Ref. 19)

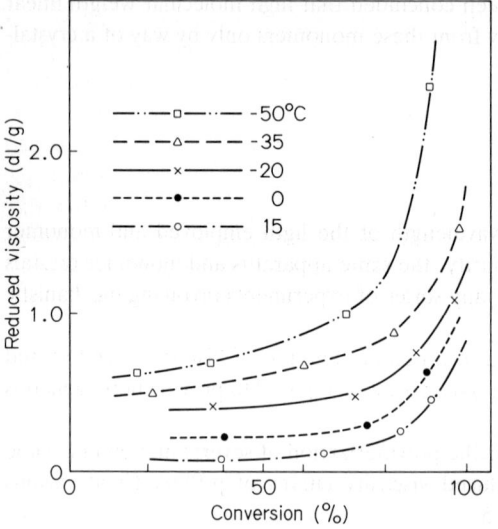

Fig. 5. Plot of conversion vs. reduced viscosity for the polymerization of p-PDA Et at various temperatures. (source: Ref. 19)

Additives, e.g. initiator, for any type of chain reaction are not involved in the crystalline state polymerization. An intermittant irradiation has no appreciable effect on kinetic curves as far as the total irradiation time is the same. An induction period has not been reported except in one case (see Sect. IV.b.)[28].

The surrounding atmosphere generally does not show any significant effect on the crystalline state photoreaction, except a few cases which have been reported recently[29, 30] (see Sect. VII.).

From these kinetic features it is concluded that the crystalline state photopolymerization of diolefinic compounds substantially follows a stepwise mechanism. The reaction of DSP is the first example of a stepwise photopolymerization[20], not only in the crystalline state but also in other forms of photopolymerization.

Studies of the temperature effect are one of the most interesting subjects in crystalline state polymerization since the temperature is closely related with molecular motion under the control of the crystal lattice. The effects of temperature on the polymerization rate and molecular weight of the polymer obtained have been studied for several monomers. Among these monomers, the effects have been fully investigated in the polymerization of p-PDA Et in the temperature range from 4.2 K (liquid helium) to temperatures above the crystal melting point[19, 28].

The temperature effect on the polymerization of p-PDA Et is described in the range from −50 to 15 °C in Figs. 4 and 5. p-PDA Et, with a melting point of 100 °C (96 °C by the capillary method) and a crystal transition point of 56 °C (DSC), photopolymerizes quantitatively to a crystalline high polymer at a temperature below ca. 0 °C as is obvious in Figs. 4 and 5. However, above ca. 25 °C, a partially cross-linked amorphous polymer is obtained in poor yield.

Figure 4 shows that raising the temperature from −50 to −20 °C increases the polymerization rate somewhat and then decreases it again in the range from −20 to 15 °C. Such an inversion of the velocity curve around 0 °C is interpreted not by a real inversion of the polymerization rate with increasing temperature but by a depression of the polymer chain growth, due to a disordering of the crystal lattice by thermal vibrations during polymerization. Therefore, at the same conversion, a higher reduced viscosity is observed for the product polymerized at the lower temperature as shown in Fig. 5. The polymer yield saturates at a lower conversion for the reaction at the higher temperature above ca. 20 °C which supports the above interpretation. For instance, at 32 °C the final conversion of poly-p-PDA Et is 67%, and the degree of polymerization is 3.4. At 45 °C the final conversion is only 18%, and the degree of polymerization is 2.3. In contrast, even at −50 °C the polymerization proceeds rather rapidly to produce a linear high molecular weight polymer in quantitative yield.

Gerasimov et al. have reported that poly-p-PDA Et is obtained quantitatively at 170 − 4.2 K and that the activation energy is 1600 ± 300 cal/mol at 170 − 100 K and close to zero (<20 cal/mol) at 90 − 4.2 K, respectively. From the outstanding reactivity of p-PDA Et at an extremely low temperature, the barrier to the reaction in the monomer crystals has been attributed to the force of the crystal lattice and classified into the region of negative values of the potential energy. In addition the observed induction period at 4.2 K has been attributed to the growth period of crystal defects (see Sect. IV.a.) In the case of DSP, quantitative conversion of monomer to polymer crystals has been achieved by photoirradiation at −60 °C[26].

As is the case of DSP and p-PDA Et crystals, most of diolefinic crystals are converted into the corresponding polymer crystals more rapidly at lower temperatures. For example, on photoirradiation of p-PDA Me crystals (m.p. 171 °C) at 3 and 19 °C for 30 min, the intrinsic viscosities of the final polymer are 4.5 and 2.7 with yields of 80 and 68%, respectively[31].

These results indicate that the polymerization readily affords a highly crystalline linear polymer in high yield as long as the polymerization temperature is sufficiently low to maintain the rigidity of the crystal lattice of the reactant. Since the monomer molecules are aligned in photoreactive crystals in such a manner that the probability of effective collisions is close to unity (see Sect. IV.a.), a concentration term in the usual sense is not definable, and the reaction velocity is prodominantly influenced by the number of rigidly fixed intermolecular double bonds in the crystals per unit volume. Such a temperature effect is common to most monomers and is one of the most remarkable features of four-center photopolymerization in the crystalline state.

Studies of the polymerization behaviour of p-PDA Et have been extended to the phase-transition temperature in the monomer crystals (56 °C) and temperatures above the monomer melting point[19]. All the results are readily explained in terms of thermal movements of the molecules in photopolymerizable crystals. At a temperature below ca. 20 °C, p-PDA Et behaves as a typical polymerizable monomer of a four-center photopolymerization. Upon irradiation at 62 °C for 20 h, a white powder is obtained at 40% conversion, and the product includes a large amount of dimer (30% based on the starting monomer, m.p. 136 °C) (Fig. 3, Sect. II.b.). This implies that at 62 °C the monomer crystal lattice begins to perform such violent molecular motions at the lattice point that the monomer-lattice controlled polymerization scarcely proceeds and is only limited to oligomer formation, including a considerable amount of dimer. At 110 °C an insoluble white powder is produced in 6% yield with a large amount of monomer and a small amount of a cross-linked product (1%). The same product is obtained from p-PDA Et by radical polymerization using an initiator (BPO) in solution at reflux temperature. Therefore, at 110 °C a vinyl-type polymerization proceeds to give a considerable amount of cross-linked substances at a later stage in the reaction.

In contrast, due to the typical temperature effect on the lattice-controlled process of a four-center photopolymerization, in the case of a few diolefin crystals such as m-PDA Me (m.p. 138 °C), only the amorphous oligomer is produced at all the temperature ranges attempted. In the polymerization of m-PDA Me higher temperatures favor chain growth. This behavior is reasonably well explained by lattice-controlled dimerization followed by random cyclobutane formation yielding the oligomer through the thermal diffusion process (Sect. IV.b.)[22].

b. Effect of Wavelength of Exciting Light

All the photopolymerizable monomers discussed in this review contain two conjugated olefinic double bonds. Since the π-electron conjugation of the monomer (A) is interrupted by the formation of a cyclobutane ring to produce a molecule larger than a dimer (B), the π-π^* electronic transition of B is shifted to a higher energy level than that of A.

A: R\R'C=CH–Ar–CH=C\R'/R

B: R\R'C=CH–Ar–[CH(C(R)(R'))(C(R')(R))CH–Ar–]$_{n\geq 1}$CH=C\R'/R

For example,

DSP : Ar = [pyrazinyl], R = [phenyl], R' = H

p–PDA Me : Ar = [phenyl], R = $-COOCH_3$, R' = H

p–CPA Me : Ar = [phenyl], R = $-COOCH_3$, R' = CN

The reaction scheme is as shown in Eqs. (1)–(6)[32].

$$A \xrightarrow{h\nu_1} A^* \qquad (1)$$

$$B \xrightarrow{h\nu_2} B^* \qquad (2)$$

$$A^* + A \xrightarrow{k_1} B \qquad (3)$$

$$A^* + B \xrightarrow{k_2} B \qquad (4)$$

$$B^* + A \rightarrow B \qquad (5)$$

$$B^* + B \rightarrow B \qquad (6)$$

In the above reaction scheme, A^* and B^* represent the species A and B, respectively, in the excited state.

Equation (3) represents a dimerization reaction and Eqs, (4)–(6) represent growth reactions. All the steps (1)–(6) proceed by irradiation with a xenon lamp or a high-pressure mercury lamp. On the other hand, only reactions (1), (3), and (4) give the oligomer exclusively upon irradiation at the long wavelength edge of the monomer A ($h\nu_1$). On successive irradiation of the oligomer with the wavelength of light which excites B ($h\nu_2$), a high molecular weight polymer is produced by step (6) which is a growth reaction of the terminal unit in the growing chain. By such a selective excitation technique, step (5) actually does not occur because all the monomer molecules (A) have already been exhausted in the course of oligomerization.

The UV absorption of crystalline DSP as measured in a thin layer deposition on a quartz plate is shown in Fig. 6.

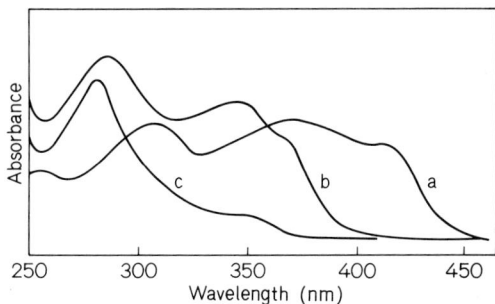

Fig. 6a–c. UV spectra of crystalline DSP (**a**), DSP-oligomer (**b**), and poly-DSP (**c**). (source: Ref. 32)

The absorption band at 430 nm is seen as a shoulder of the stronger band at 370 nm. Upon irradiation of crystalline DSP with 430 nm monochromatic light, the UV absorption spectrum changes, and new peaks appear at 350 nm and 290 nm (Fig. 6). The IR absorption band at 970 cm^{-1} corresponding to ν_{CH} of a trans HC=CH bond gradually decreases, and then reaches a constant level at about one fifth of the initial values (Fig. 7). At this stage, the monomer in KBr pellets has almost vanished.

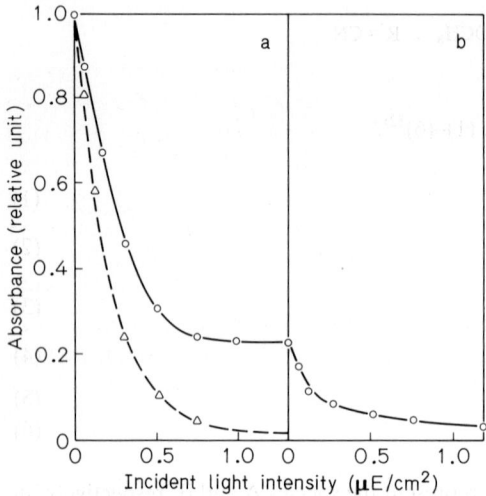

Fig. 7a, b. Polymerization of DSP in KBr irradiated at (**a**) 430 nm and (**b**) 350 nm. The plots, -o- and --△--, show the amounts of double bond and monomer, respectively. (source: Ref. 32)

A powdery white product is obtained quantitatively from DSP and melts at 285–295 °C with decomposition. Assuming that the coefficient of absorptivity of the product at 970 cm^{-1} is the same as that of the monomer, the IR absorption of the product corresponds approximately to that of a pentamer (Fig. 7).

The NMR spectrum of the product has peaks corresponding to cyclobutane protons near $\pi = 5$ and olefinic and aromatic protons at $\pi = 2$–3. The degree of polymerization calculated from the ratio of the number of cyclobutane protons to that of the other protons is equal to about one-fifth. This value corresponds to a pentamer which is in good agreement with that estimated by IR measurements. On further irradiation of DSP oligomers at wavelength longer than 400 nm, no further change is observed. However, the oligomer is converted into a high polymer by exposure to light of wavelengths shorter than 400 nm. In the DSP oligomer excitation at 350 nm induces a decrease in the UV absorption band at 350 nm, leaving the one at 290 nm unchanged (Fig. 6). At the same time, the IR absorbancy at 970^{-1} cm decreases and then completely disappears (Fig. 7). When DSP crystals are irradiated at 350 nm, the UV and IR absorption spectra gradually change into those of the polymer apparently without ever exhibiting the spectral behavior of the oligomer during the process.

From these results it is concluded that upon irradiation at wavelengths longer than 400 nm, crystalline DSP is not converted into a high polymer but quantitatively into the oligomer (pentamer on average) and that wavelengths shorter than 400 nm are required for further growth of the terminal unit in the oligomer[33].

Other monomers, e.g. p-PDA Me and p-CPA nPr, have also been oligomerized in the crystalline state upon irradiation at the long wavelength edge of the absorption, and each oligomer has been polymerized to a high polymer upon further irradiation with a shorter wavelength of light[32]. Here, the average molecular weight of the oligomer is determined by the ratio of the rate constants of Eqs. (3) and (4). For example, the pentamer formation of DSP corresponds to the case where k_2 is about ten times larger than k_1, suggesting the acceleration effect in the course of polymerization. The reason why k_2 is larger than k_1 cannot be explained satisfactorily.

The polymer prepared by selective excitation of the reactive species is essentially the same as that produced by direct polymerization with excitation of the monomer and the terminal olefin. It should be emphasized, however, that successive polymerization of oligo-p-CPA nPr does not proceed completely whereas the direct polymerization of the monomer, p-CPA nPr, by irradiation with a xenon or a high-pressure mercury lamp quantitatively yields polymer crystals of higher molecular weight[34]. Such differences of polymer chain growth, which depend on the photoexcitation method, are observed also in the polymerization of DSP. In DSP, the reacting crystals expand in the course of selective oligomerization (1.244 → 1.18 g/cm^3), and with subsequent polymerization of the oligomer, the crystals start to shrink (DSP oligomer crystal 1.18 g/cm^3, poly-DSP crystal 1.257 g/cm^3)[26]. The crystal density change during the whole polymerization of DSP suggests that, compared to the two-step polymerization, the overall volume change of the reacting crystals is considerably suppressed in the polymerization by an ordinary light source. Consequently, disordering of the crystal will be minimized during polymerization.

It is noteworthy that a harmonious combination of both visible and ultraviolet light is necessary to complete the growth of polymer chains. At the same time, the effect of wavelength of the exciting light on the four-center photopolymerization confirms the stepwise mechanism as suggested by kinetic studies (see Sect. III.a.). In this type of photopolymerization, the molecular weight of the resulting polymer may be easily controlled by means of reaction temperature and/or wavelength of the light.

c. Quantum Yield

The quantum yields of the oligomerization and polymerization of DSP, P2VB and p-PDA Me have been measured with respect to the number of olefinic double bonds consumed to form cyclobutane per absorbed quantum[32]. The quantum yield (Φ) is defined by the equation $\Phi = (dc/dt)/I_{abs}$ where dc/dt is the rate of disappearance of the olefinic double bonds per unit volume and I_{abs} the rate at which the incident light is absorbed per unit volume of the KBr pellet containing the sample. The rates of disappearance of the olefinic double bonds during oligomerization and polymerization were monitored by IR spectroscopy. In the case of oligomerization, KBr pellets containing monomer crystals were photoirradiated. In the polymerization of the oligomer before the measurement, the KBr pellets with the monomer crystals were exposed to light at the wavelength edge of the absorption of the monomer until the monomer has been completely converted into the oligomer.

The initial quantum yields for the oligomerization and the polymerization of DSP, P2VB, and p-PDA Me are summarized in Table 2.

Table 2. Quantum yields of oligomerization and subsequent polymerization of DSP, P 2 VB, and p-PDA Me[32]

	Wavelength used in irradiation (oligomerization) (nm)	Φ	Wavelength used in irradiation (polymerization) (nm)	Φ
DSP	436	1.2	365	1.6
P 2 VB	405	0.04	–	–
p-PDA Me	365	1.2	313	0.7

Since the scattered light is included in the absorbed light, the quantum yields may be expected to be higher than those listed in Table 2. The quantum yields of oligomerization and polymerization of DSP and p-PDA Me are between 0.7 and 1.6.

Holm and Zienty have measured the quantum yield for the overall polymerization process of α, α'-bis(4-acetoxy-3-methoxybenzylidene)-p-benzenediacetonitrile (AMBBA) crystals in slurries where scattered light was calibrated and the conversion determined by UV and IR spectroscopy or by the weight of isolated solid residues[7]. The quantum yield of the polymerization of AMBBA is 0.7 which was determined on the basis of the disappearance of two double bonds (1.4 if assigned on the basis of the number of double bonds disappeared).

These quantum yields indicate that these photoreactions belong to single-photon reactions in which the theoretical maximum value is equal to 2, and that some of these reactions proceed very efficiently. Such a high quantum yield may reflect a high probability of effective collisions in the four-center photopolymerization in the crystalline state.

Higuchi et al. have shown that the reactivity of DSP and P 2 VB crystals is explained by the stabilization energy of the transient complex which, in turn, depends on both the electronic structure of the monomer molecule and the intermolecular arrangement in the monomer crystal[35].

Different quantum yields of these reactions are not explained satisfactorily though overlap of electronic orbitals in the monomer crystal, changes of crystal volume, changes of bond angle or displacement of the center of gravity of the monomer unit during the reaction, growth direction of the polymer chain, etc. should be closely related to the reactivity.

d. Matrix Effect. Solution Photopolymerization and Photodepolymerization

For the purpose of visualizing the role of the crystal lattice ("matrix effect") during the reaction, a comparison of the chemical behavior in solution and in the molten state is meaningful. Photoinitiated vinyl-type polymerization has been observed with p-PDA Et in the molten state (see Sect. III.a.)[19]. Photopolymerizability in solution has been investigated with DSP series, p-PDA Et, and p-CPA nPr, including non-photopolymerizable compounds in the crystalline state[34, 36–39].

As a result, all these diolefins except p-CPA nPr have been found to be converted into their oligomers in solution upon irradiation at the long wavelength edge of the absorption which excites only the monomer or monomer aggregates and not the terminal olefinic bond in the oligomer. For example[36], when 0.2 g of DSP in 50 ml of tetrahydrofuran is irradiated with light of wavelength longer than 380 nm at room temperature for 48 h, a powdery substance is isolated in 60% yield. The product has a molecular weight of 900, and an oligomer structure for DSP having a cyclobutane ring in the main chain has been confirmed by NMR and IR spectoscopy. However, the NMR peak due to the hydrogen atoms attached to the cyclobutane ring in the DSP oligomer is not as sharp as the corresponding peak in the oligomer prepared in the crystalline state. The DSP oligomer prepared by solution photopolymerization is amorphous according to X-ray powder analysis and, in addition, the amorphous solid oligomer does not grow into a high molecular weigth polymer when irradiated at 380 nm as is true of the as-prepared crystalline oligomer. Such differences between the two kinds of oligomers indicate that the oligomer obtained by way of solution polymerization contains mixed structures of cyclobutane units with various steric configurations.

Diolefinic compounds other than DSP such as P 2 VB, P 3 VB, P 4 VB, 1,4-distyrylbenzene (DSB), and p-PDA Et, are oligomerized in solution in a way similar to DSP. Exciting wavelength, yields, and average molecular weight of the resulting oligomers are compiled in Table 3.

Table 3. Solution oligomerization of diolefinic compounds[36, 37]

	Irradiation (nm)	Yield (%)	Molecular weight
DSP	> 380	60	770–900
DSB	> 360	100	634
P 2 VB	> 360	29	814
P 3 VB	> 360	94	1074
P 4 VB	> 360	73	682
p-PDA Et	> 340	30	650

Reaction conditions: initial concentrations = 0.2 g monomer/50 ml THF (except p-PDA Et), 10.0 g monomer/50 ml acetonitrile for p-PDA Et. Reaction time; 48 h (except p-PDA Et), 12 h for p-PDA Et. Reaction temperature: room temperature

Table 3 shows that crystalline DSP, P 2 VB, and p-PDA Et are photopolymerizable whereas the other three compounds are photostable in the crystalline state.

In solution, as the unimolecular cis-trans isomerization of excited species seems to compete with photocycloaddition polymerization, a highly concentrated solution of the monomer is advantageous for oligomerization. Such prominent difference of reactivities suggests extremely high stereoselectivity due to the crystal lattice in the crystalline-state reaction.

Further studies on the the solution behavior have been carried out on p-PDA Et[37–39]. In a concentrated solution of p-PDA Et in acetonitrile, the photoproduct distribution varies with irradiation wavelength[37], that is on irradiation at > 300 nm the photoreaction results in a diversity of products due to the excitation of the complexes and isolated p-PDA Et molecules[38]. On the other hand, the distribution is controlled by irradiation with ligth of wavelength > 340 nm that excites only the ground-state complexes. These complexes are only formed in highly concentrated monomer solutions and facilitate p-PDA Et excimer formation in the excited state via which stereoselective photodimeriza-

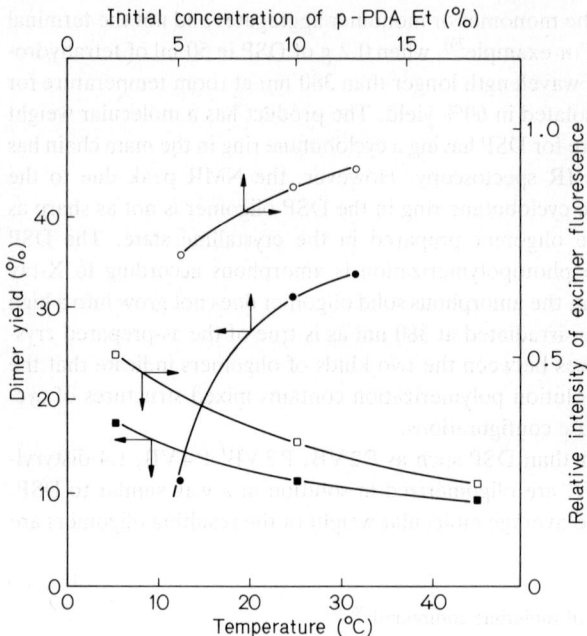

Fig. 8. Concentration and temperature dependence of excimer fluorescence and 1,2-dimer yield of p-PDA Et upon photoirradiation in solution. (source: Ref. 39)

tion preferentially occurs to give the head-to-head type dimer (1,2-dimer) with a mirror symmetry of the cyclobutane ring. The structure of the 1,2-dimer has been determined by elemental analysis, IR, MS, and NMR spectra[39].

Photodimerization via excimer formation has not only been confirmed by quantitative spectral and photochemical data but also by determining the dependence of excimer fluorescence and photodimer yield on the initial p-PDA Et concentration and temperature (Fig. 8).

The molecular structure of the 1,2-dimer indicates an orientation of a pair p-PDA Et molecules parallel to the head-to-head direction along the molecular axis.

As is obvious in Fig. 8, the p-PDA Et dimer with the mirror symmetry of cyclobutane is obtained in appreciable amounts in highly concentrated monomer solution and at low temperature upon irradiation with light of wavelengths > 340 nm.

In contrast to the high photopolymerizability of the crystalline oligomer (as-prepared), the recrystallized oligomer or the oligomer prepared in solution do not show further photopolymerizability.

Thus, the reaction of the excited double bonds in the oligomer may follow two different paths which are entirely opposite to each other, cyclobutane formation and cleavage. Cyclobutane formation is favored by the molecular arrangement in photopolymerizable crystals while cleavage of cyclobutane ring becomes predominant in solution.

This is a striking example of a matrix effect originating in the topochemical process of a substance which is chemically the same (oligomer) but behaves in three different ways: photodepolymerization, photopolymerization, and no photoreaction, depending on its physical state (in solution, in as-prepared and recrystallized crystalline states)[33]. Further discussions on the matrix effect have been made in correlation with crystallographic studies (see Sect. IV.b. and VII.).

IV. Crystallography

From studies on the polymerization behavior of four-center photopolymerizations in the crystalline state, several remarkable features have been found which suggest significant participation of the crystal lattice in polymerization. For example, the plate-like crystals of DSP (α) obtained by recrystallization from solution photopolymerize in the crystalline state whereas DSP crystals (γ) obtained through sublimation are needle-like and do not undergo any photochemical change[20]. Moreover, plate-like crystals of DSP and P2VB, and crystals of poly-DSP and poly-P2VB respectively are very similar to each other in their major X-ray diffraction peaks (Fig. 9)[8, 40]. X-Ray analyses of the crystal structures of several diolefinic compounds, including photostable DSP (γ) crystals, have been performed mostly by Nakanishi and Sasada[26, 41–51].

An empirical rule on a similar photoreactivity between mono- and di-olefin crystals (see Sect. II.a.) has been explained by correlating common molecular shape and packing with common intermolecular contacts between electron-rich and electron-deficient moieties in these olefin crystals.

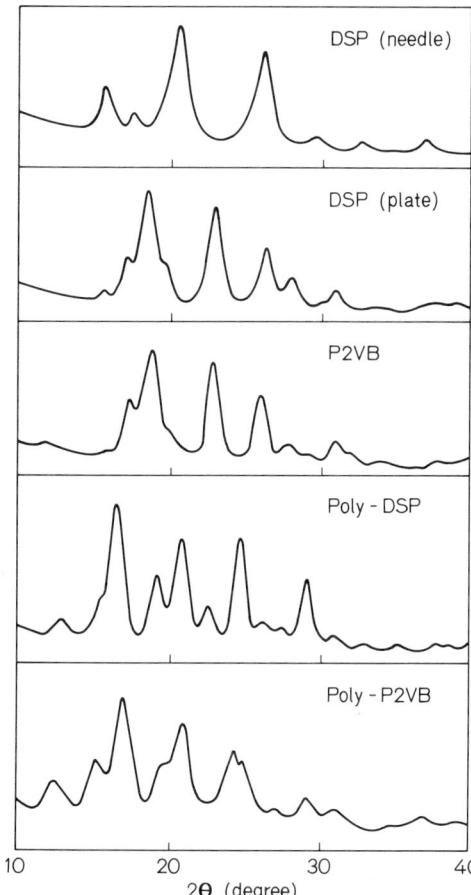

Fig. 9. X-ray diffraction diagrams of DSP (γ-form, needles), DSP (α-form, plates), P2VB, poly-DSP, and poly-P2VB (source: Ref. 8)

Changes in the crystal structure during polymerization have also been pursued by X-ray crystallographic techniques[26, 44]. The results indicate that in all photopolymerizable crystals the molecules are arranged such that the relative positions and orientations of the double bonds are favorable for cyclobutane formation. Further, the principal features of the molecular arrangement in the monomer crystals remain unaltered in the polymer and oligomer crystals. All the polymer crystals are three-dimensionally oriented. In addition, a 1,3-trans configuration, which had been assumed from NMR spectroscopy with respect to the cyclobutane ring of all the polymers, was concluded.

Full details of crystallographic studies are described a in a reference[44].

a. Molecular Alignment and Relative Orientation of Monomer and Polymer Crystals

From the viewpoint of the topochemical process, the diolefinic compounds prepared so far are classified into three groups: photopolymerizable, photooligomerizable[1], and photostable crystals. Among these, nine kinds of monomer crystals and their polymers have been subjected to structure analysis so far. The obtained crystal data are compiled in Table 4.

The crystal structures of cinnamic acid (α), DSP (α), P 2 VB, and p-PDA Me are schematically described in Fig. 10.

As is obvious in Fig. 10, there are common features of molecular packing in photoreactive crystals. In all the photopolymerizable crystals in Table 4, nearly planar molecules are piled up and displaced in the direction of the molecular longitudinal axis by about half a molecule to form a parallel plane-to-plane stack. The periodicity in the stack is about 7 Å. The shortest intermolecular distance between the double bonds in photopolymerizable crystals is about 3.9 Å (Table 4) and it is found between molecules related to the center of symmetry in the stack. The second shortest distance between molecules in different stacks is more than 5 Å. Therefore, the double bonds in the stack react to form a cyclobutane ring; consequently, polymer chains should grow in the direction of the stack. The crystal axis along the stack in each photopolymerizable crystal, i.e. the presumed chain-growth direction, is indicated by c) in Table 4.

Figure 10 reveals that molecular shape and packing of photoreactive crystals are very similar. In addition, the electron-rich nitrogen or carbonyl group approaches the electron-deficient benzene ring in all the photopolymerizable crystals[41, 42, 44, 46, 49, 50]. These two common features may govern the formation of photopolymerizable crystals and thus allow an empirical rule of similar photoreactivity between mono- and di-olefin crystals to be established (see Sect. II.a.).

From the oscillations and Weissenberg photographs, all the polymers have been found to be oriented three-dimensionally, while oligomers of m-PDA Me and CVCC Me are amorphous.

The periodicity in the polymer chain with alternating cyclobutane and 1,4-arylene units is between 7.50 and 8.16 Å. Moreover, according to a model construction (Fig. 11) which shows an especially good fit between a polymer chain and a stack of monomers,

---

1 The photooligomerizable crystal in this chapter is the diolefin monomer crystal which is converted only into an amorphous oligomer

Table 4. Crystallographic data of diolefinic monomers and their polymers

	Compound	Space group	a (α)	b (β)	c (γ)	Z	Dx	C–C Distance[a]	Reactivity[b]	Ref.
DSP (α)	Monomer	Pbca	20.638	9.599	7.655[c]	4	1.244	3.939	P (high)	48
	Polymer	Pbca	18.36	10.88	7.52[d]	4	1.257	–	–	40
P2VB	Monomer	Pbca	21.060	9.567	7.311[c]	4	1.281	3.910	P (low)	49
	Polymer	Pbca	18.9	10.5	7.53[d]	4	1.26	–	–	50
p-PDAMe	Monomer	P$\bar{1}$	7.190[c] (95.08)	8.404 (117.06)	5.883 (78.00)	1	1.319	3.957	P (high)	55
	Polymer	P$\bar{1}$	7.82[d] (107.8)	7.42 (106.0)	6.04 (78.8)	1	1.29	–	–	50
p-PDAEt	Monomer	P2$_1$/a	7.399[c]	9.894 (99.74)	10.167	2	1.242	3.970	P (medium)	47
	Polymer at −15°C	P2$_1$/a	8.16[d]	9.98 (102.0)	8.62	2	1.30	–	–	50
p-PDAPh	Monomer	P2$_1$/c	6.917[c]	18.584 (101.87)	7.557	2	1.293	3.917	P (medium)	56
	Polymer	P2$_1$/c	7.50[d]	17.3 (102.0)	7.50	2	1.29	–	–	50
p-CPAnPr	Monomer	P2$_1$/n (P2$_1$)	5.341	26.112 (103.81)	6.822[c]	2	1.265	3.931	P (high)	52
	Polymer	P2$_1$/n (P2$_1$)	6.19	20.1 (96.0)	7.58[d]	2	1.25	–	–	50
m-PDAMe	Monomer	Pmn2$_1$	26.419	3.960[c]	5.935	2	1.318	3.960	O (medium)	51
CVCCMe	Monomer	P2$_1$/a	11.387	29.737	3.956[c]	4	1.346	3.956	O (high)	54
DSP (γ)	Monomer	P2$_1$/a	13.833	18.615 (92.63)	5.823	4	1.261	4.187 / 4.369	N	53

[a] Intermolecular distance between reactive double bonds
[b] P, O, and N denote polymerization, oligomerization, and no reaction, respectively
[c] Direction of chain growth
[d] Direction of polymer chain

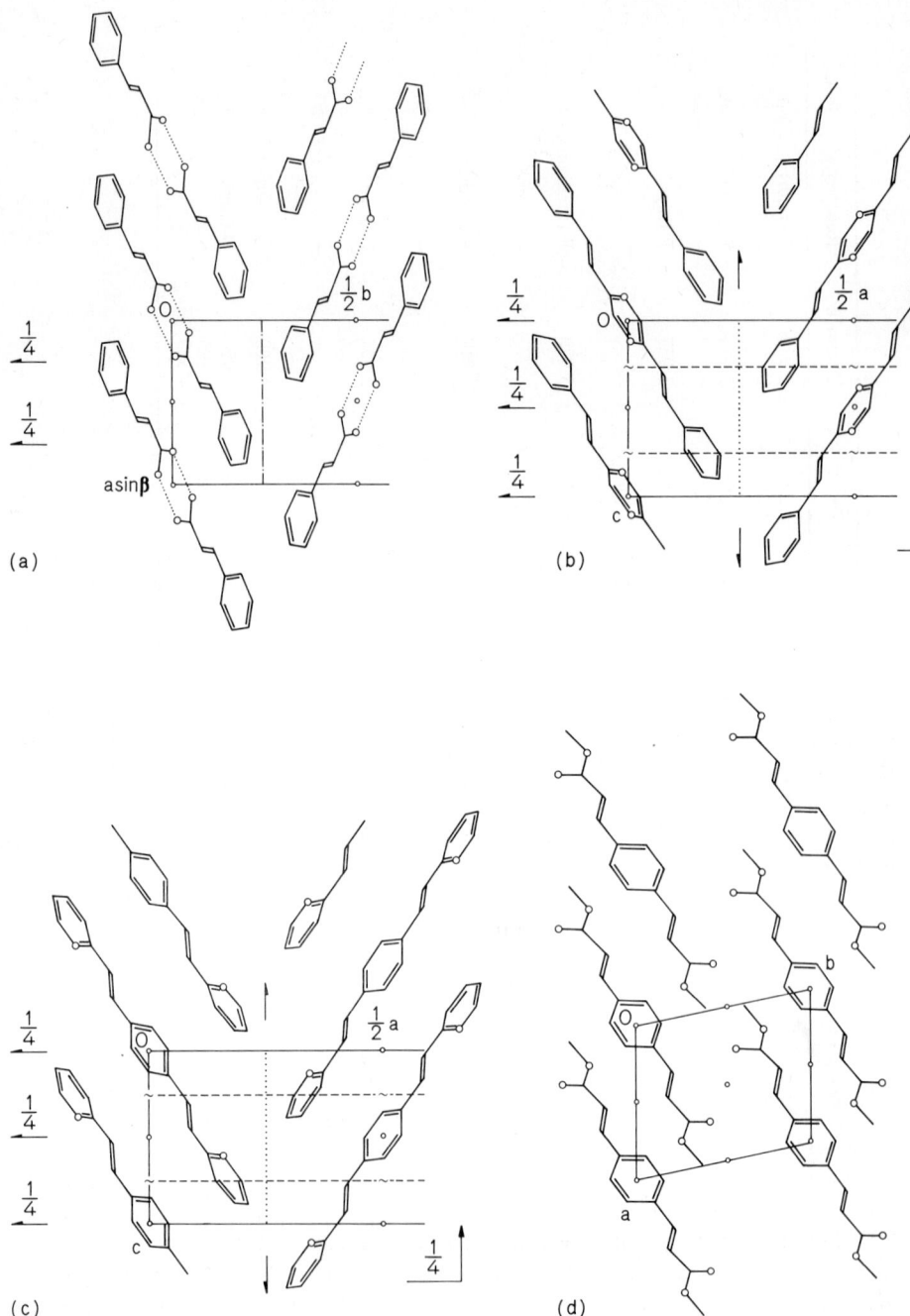

Fig. 10a–d. Crystal structures of (**a**) α-trans cinnamic acid, (**b**) DSP (α), (**c**) P 2 VB, and (**d**) p-PDA Me. (source: Ref. 11 and 44)

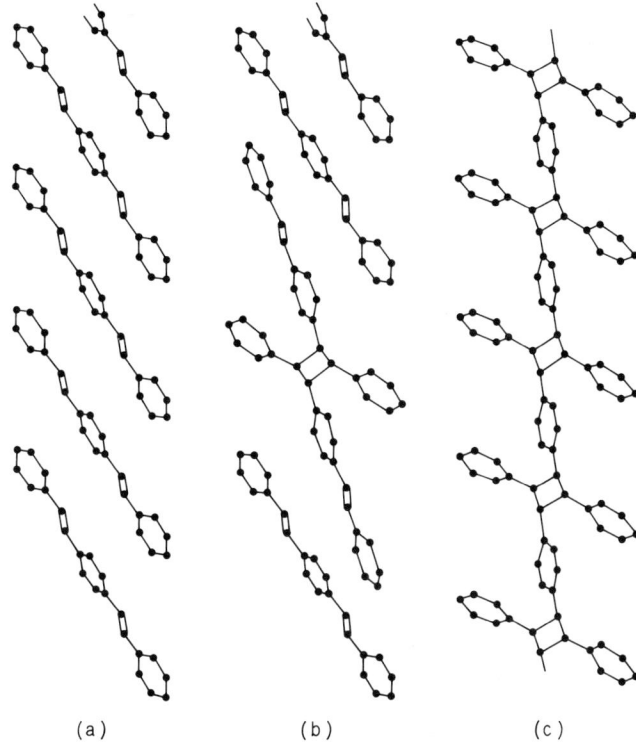

Fig. 11 a–c. Schematic illustration of the conversion of (**a**) monomer DSP (α) into (**b**) dimer and (**c**) polymer. (source: Ref. 11 and 44)

the polymer chain axis would be near the presumed chain growth direction of the monomer crystal. Thus, the polymer chain axis has been definitely assigned as indicated by d) in Table 4.

Since the olefinic double bonds in the monomer crystals are related to the center of symmetry, the 1,3-trans configuration, which had been presumed by NMR spectroscopy, was confirmed to exist in all the polymers except in those prepared from m-PDA Me and CVCC Me (Table 4). Furthermore, 1,3-trans configuration is concluded from the space groups of the polymers since these groups absolutely require that the center of symmetry is in the center of the planar cyclobutane ring.

The relative orientation of monomer and polymer crystals was determined by rotating single crystals of photopolymerizable diolefins on photoirradiation.

In the DSP (α) and P 2 VB crystals, the directions of three crystal axes of the polymer coincide with those of the monomer[43]. The relative orientation of P 2 VB and poly-P 2 VB crystals has been independently reported by Baughman[52].

A different type of relative orientation is seen in p-PDA Ph, p-PDA Et, and p-CPA nPr[44]. In these cases, the direction of the unique axis (b-axis) of the polymer coincides with that of the monomer while the directions of the other two axes do not.

In the case of p-PDA Me, the directions of all axes of the polymer do not coincide with those of the monomer. However, the temperature effect on the reaction behavior and the continuous changes of the X-ray diffraction diagrams of all monomers except m-

PDA Me, CVCC Me and DSP (γ) in Table 4 demonstrate that the polymerization proceeds by a diffusionless crystal-lattice controlled mechanism[31].

In the photooligomerizable crystals of m-PDA Me and CVCC Me[48], nearly planar molecules are piled up and overlap completely along the shortest crystal axis of about 4 Å. In the photostable crystalline DSP (γ)[47], molecules form a characteristic layer-type packing without any overlap of adjacent molecules. The γ-form of DSP has been shown to undergo an unusual, thermally stimulated, topotactic phase transformation into the α-form[53].

In p-PDA Et, a crystal transition is observed at 56 °C by DSC measurement. Packing modes of the crystals above and below 56 °C are nearly identical with only small differences in the orientation of the terminal ethoxycarbonyl groups. This explains why the crystal transition of p-PDA Et scarcely affects the polymerization behavior[51].

Excimer fluorescence has been reported on DSP (α) and CVCC Me crystals[54]. Further studies are required to correlate the emission spectrum with the crystal structure of these monomers.

b. Polymerization Mechanism Based on Topotaxy

In order to visualize the details of the elementary processes in polymerization, the monomer crystal-lattice control of the three processes – initiation (i), propagation (ii), and crystallization of polymer (iii) – has been examined on the basis of structural characteristics of the resultant polymer[44].

The monomer lattice control of process (i) is verified by the reactivity difference of chemically related monomers with different crystal structures such as DSP (α) and (γ) crystals. It is also operative for process (ii) of the monomers as deduced from the fact that the cyclobutane ring in the polymer chain retains the same symmetry as the olefinic bonds in the monomer crystal. In the photocycloaddition of p-PDA Et in solution, cyclobutane rings with different symmetries (m and l) have been found[37–39]. The control of process (iii) is evident from the orientation of polymer crystals and from the preservation of the crystal symmetry during polymerization (Table 5). It is noteworthy that the crystal system and space group of the monomer are precisely transmitted to the polymer in all the cases examined although there are three different kinds of crystal systems. This is in remarkable contrast to the topotactic polymerization of cyclic oligomers of formaldehyde[55]. Thus, the monomer lattice-control of the whole processes is generally established for the four-center photopolymerization of rod-like diolefinic compounds.

Topotaxies found in the polymerization are classified into three groups according to the degree of topotactic control, relating the coincidence of the space groups to the directions of three axes of the monomer and polymer crystals[44].

In the course of the polymerization of DSP (α), the a-axis is contracted by 11% and the b-axis is elongated by 13% whereas the c-axis is contracted only by less than 2%. In order for the reaction to occur between the intermolecular double bonds of the diolefin (Fig. 11), these double bonds must approach each other from the van der Waals separation distance (3.939 Å) to the bond length of cyclobutane ring (1.56 Å). However, the center of the pyrazine ring which is the center of gravity of the monomer unit does not move as much along the c-axis. The approach of the first two double bonds results in a close contact of another double bond in the same molecule with the double bond of an

Table 5. Topotaxies in the four-center photopolymerization of diolefinic monomers in the crystalline state[44]

Group	Coincidence of crystal symmetry between monomer and polymer	Monomer	Crystal system	Space group
1	Crystal system, space group and directions of three axes	DSP (α) P2VB	Orthorhombic Orthorhombic	Pbca Pbca
2	Crystal system, space group and direction of unit axis	p-PDAEt p-PDAPh p-CPAnPr	Monoclinic Monoclinic Monoclinic	$P2_1/a$ $P2_1/c$ $P2_1/n$ and $P2_1$
3	Crystal system and space group	p-PDAMe	Triclinic	$P\bar{1}$

adjacent molecule (Fig. 11). Thus, the center of gravity of the rigid molecule with two functional groups may act as the fulcrum of a seesaw during the chain growth. A similar polymerization mechanism has been proposed in terms of shearing polymerization for thermally or photochemically induced crystalline-state polymerization of diacetylenes[56].

In conclusion, the four-center photopolymerization is a novel type of topochemical reaction which is crystal-lattice controlled with respect to the whole set of elementary processes[44] including initiation, propagation and crystallization of polymer.

In a recent report[57] it has been stated that, according to X-ray diffraction analysis and electron microscopy, the space group of DSP (Pbca) is not maintained but that the polymer phase has space group $P2_{1ca}$ and the four-center photopolymerization of DSP occurs along the lines of a heterogeneous solid-state reaction which starts at defect sites within the crystal. Thus, the structure of the product is predefined by the arrangement of the molecules around the particular defect. Gerasimov et al. have also suggested a heterogeneous process for the photopolymerization of p-PDA Et crystals assuming a growth of crystal defects during the induction period[28].

In contrast to the above report, Nakanishi et al. have reconfirmed the previous conclusion by the Buerger precession technique[58] that the space group of poly-DSP is Pbca, and consequently the 1,3-trans configuration with respect to the cyclobutane ring in the main chain of the polymer is uniquely derived. Jones[59] has utilized transmission electron microscopy and "real space crystallography"[59,60] to observe single crystals of DSP before and after polymerization. As a result, they have supported the homogeneous mechanism of the reaction of DSP.

The successesful asymmetric synthesis by Lahav et al.[9] also supports the diffusionless mechanism of the four-center photopolymerization since the direct rearrangement controlled by the monomer crystal to the polymer crystal is predetermined by the crystal structure design of the monomer ("crystal engineering").

The polymerization of P2VB, which belongs to group 1 in Table 5, also involves such direct rearrangement although the displacement of a monomeric unit in the direction of chain growth results in an elongation of 3.0% in the direction of the c-axis with crystal expansion, in contrast to the contraction of 1.8% with crystal shrinkage of DSP (α)[43]. A similar mechanism was suggested for the polymerization of α, α'-bis-(4-acetoxy-3-methoxybenzylidene)-p-benzenediacetonitrile[7].

From the viewpoint of direct rearrangement, polymerizations of other diolefin crystals are of interest; the growth direction of the polymer chain deviates from the expected direction of chain growth in the monomer crystal. The movements of the monomeric units in these compounds, which can be estimated from the periodicities in the monomer in the chain growth direction and along the polymer chain axis, are comparatively large (e.g. elongation of 8.4% for p-PDA Ph, 8.8% for p-PDA Me and 11.1% for p-CPA nPr). On the other hand, no significant density changes (less than 4%) are observed in the polymerization of all the monomers. Therefore, it is conclusive that if the displacement of monomeric units in the chain growth is appreciable, accumulation of strain energy results in an impairment of the crystals which leads to phase separation of the polymer-rich portion from the remaining monomer crystals. Such successive changes proceed from the surface to the interior of the monomer crystal.

In the case of diolefinic compounds which polymerize only to oligomer such as m-PDA Me crystals the reaction mechanism is quite different[22]. At the initial stage of photoirradiation, the molecules of m-PDA Me related by the shortest translation react topochemically to form the corresponding dimer having a cyclobutane ring with a mirror symmetry[45]. Then, the subsequent reaction between the dimer and its neighbor in the destroyed crystal lattice results in an amorphous oligomer exhibiting more than two kinds of cyclobutane structures. The rate of oligomerization from the dimer is accelerated as the temperature increases indicating that oligomerization of these compounds is more likely to occur than the reactions accompanied by thermal diffusion. Thus, the existence of two kinds of reaction mechanism, i.e. topochemical dimer formation in the regular lattice and subsequent random cycloaddition in a disordered crystal lattice, is presumed for m-PDA Me.

V. Changes of Morphology and Thermodiagram in the Course of Polymerization

It is of great interest to investigate the continuous changes of morphology and thermodiagram in the course of direct rearrangement of monomer crystals to polymer crystals and to correlate these changes in terms of crystallographic data.

Morphological changes, which are classified into four groups, have been correlated to the degree of topotactic control. Thermal analysis has been studied on DSP~poly-DSP in some detail. Two main endothermic peaks of as-polymerized poly-DSP crystals are characterized as thermal depolymerization in the crystalline state and crystal melting point followed by thermal depolymerization in the molten state. From the results of the studies on the heat treatment of as-polymerized polymer crystals, a reversible topochemical processe has been established.

a. Morphological Changes in the Polymerization

Single crystals resulting from polymerization of several diolefinic compounds have been grouped into four classes (I–IV) of morphological changes (Table 6).
Micrographs of morphology classes I, II, and IV are shown in Fig. 12 a–c[61].

Upon photoirradiation of DSP (a), which belongs to class I, regular cracks are formed in the direction of chain growth which is parallel with the c-axis of monomer and

Table 6. Classification of morphological changes in polymerization of diolefinic monomers[23, 61]

Group	Compound	Morphological change	Molecular movement during reaction
I	DSP (α), P2VB	crystalline, large cracks and changes in shape	small
II	p-CPA nPr	crystalline, fine cracks, fibrillization	large
III	p-PDA Me, Et, Ph	crystalline, no cracks but deformation	large
IV	p-PDA iPr, Oct, m-PDA Me, Pentaerythritol tetracinnamate	amorphous, no change in shape	unknown

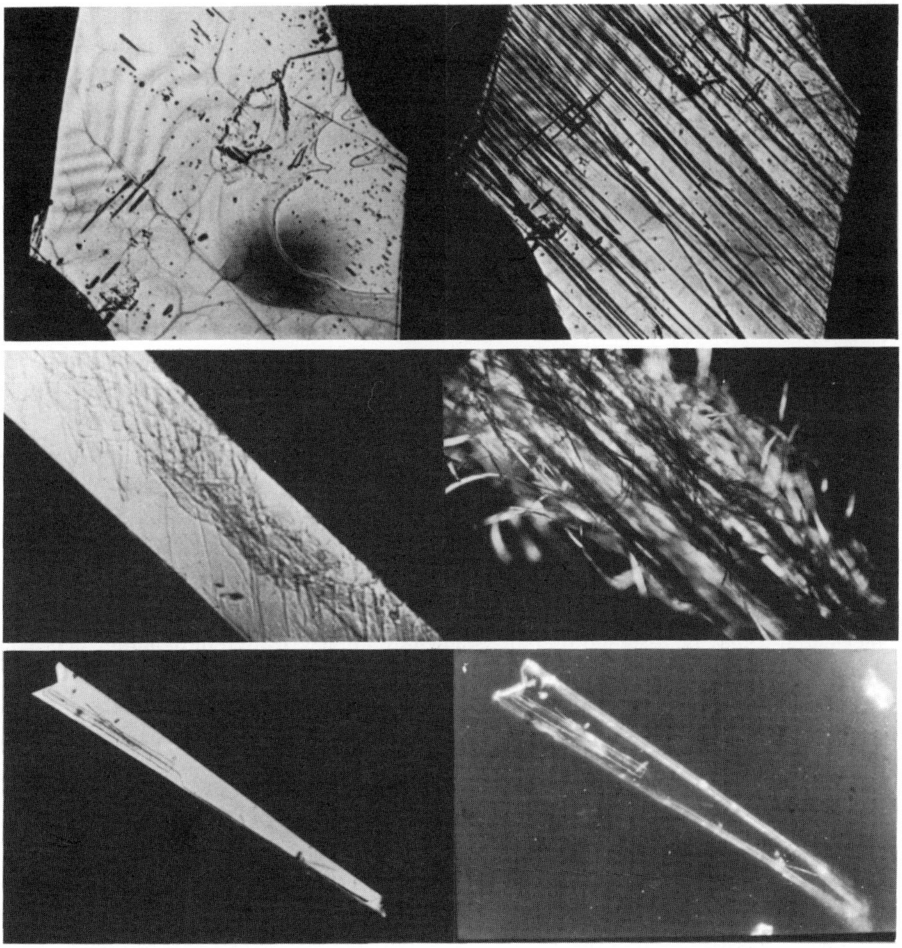

Fig. 12. (a) Micrographs of monomer and partially polymerized crystals of DSP, (b) micrographs of monomer and partially polymerized crystals of p-CPA nPr, (c) micrographs of m-PDA Me crystals and oligomer. (source: Ref. 61)

Fig. 13. Scanning electron micrograph of poly-DSP (× 10,000). (source: Ref. 62)

polymer crystals (Fig. 12). The cracks increase with irradiation time, resulting in aggregates of rod-shaped crystallites. Scanning electron microscopy of poly-DSP shows aggregates of single crystals elongated in the direction of the polymer chain (Fig. 13)[62]. Such cracking may be caused by molecular displacement in the rigid crystal lattice.

It is obvious from Fig. 12 that crack formation does not start at emergent imperfections which has been confirmed by transmission electron microscopy (see Sect. IV.b.)[59]. The scanning electron micrographs of poly-DSP and poly-P2VB[44] crystals are quite different. The difference may be reflected by a crystal volume change during polymerization in which the DSP crystals shrink while the P2VB crystals expand.

In the case of p-CPA nPr (class II single crystals), fine cracks also run in the direction of chain growth exhibiting however marked fibrillization (Fig. 12). Thus, the polymer single crystal does not retain its original monomer crystal shape. In the case of diolefin crystals of class III, fine crack formation occurs in random directions, and the final crystal deforms without breaking into pieces. Different types of morphological changes, i.e. pseudomorph formation, is observed for diolefin crystals of class IV whose polymerization is not a crystal-lattice controlled process. For example, m-PDA Me crystals gradually become amorphous without any change of the outer shape (Fig. 12c). This type of change is also seen in such crystals as p-PDA iPr[15] or pentaerythritol tetracinnamate[23].

Quite recently, Nakanishi et al. have reported an example of crystalline-state dimerization for which the product matrix is essentially of single-crystal character[63]. On the other hand, it may be assumed that any solid-state polymerization of diolefinic crystals, which results in an amorphous product, gives a pseudomorph.

From the observations of morphological changes, it can be concluded that the appearance of reacting phases is rather clearly correlated to the degree of topotactic control in the polymerization (see Sect. IV.b.).

b. Changes in the Thermodiagram during Polymerization

Figure 14 shows typical differential thermometry and thermogravimetry curves (DTA and TGA) of DSP and crystalline poly-DSP in a helium current[64].

At 321 and 339 °C, sharp endothermic peaks are observed on the DTA curve of poly-DSP, and a broad endothermal peak with a maximum at 410 °C appears as well. On the

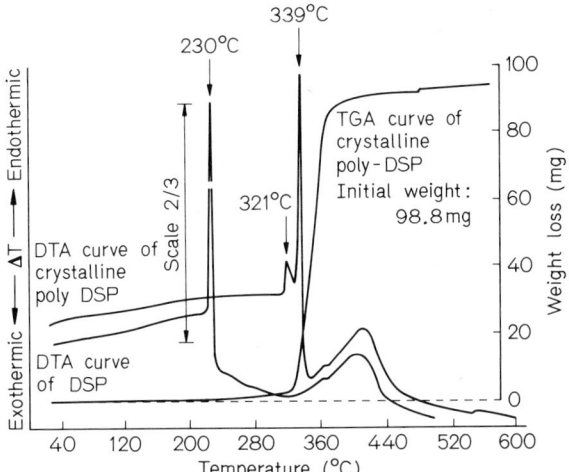

Fig. 14. DTA and TGA curves of DSP and crystalline poly-DSP in helium. (source: Ref. 64)

TGA curve, an abrupt weight loss occurs a little above 320 °C where the sample apparently becomes liquid and a significant amount of monomer crystals is detected in the quartz tube of the heating balance. This indicates sublimation of monomer formed through depolymerization. The DTA curve of the monomer (DSP) shows a melting point peak at 230 °C. Above 360 °C, the DTA curve of DSP coincides with that of poly-DSP since poly-DSP depolymerizes nearly quantitatively to DSP at high tempertures. On the other hand, the DTA curve of amorphous poly-DSP shows only a single sharp endothermic peak at 325 °C in helium.

The continuous change of the thermodiagram in polymerization has mainly been studied for DSP by means of differentially scanning calorimetry (DSC)[10]. DSC curves obtained in the course of photopolymerization of DSP crystals at the irradiation times of 20, 40, 50, and 70 min with a xenon lamp are shown in Fig. 15 together with those of DSP, as-polymerized, recrystallized, and amorphous poly-DSP's.

At the initial stage of irradiation (20 and 40 min), the DSC curve shows a melting point depression of the monomer from 226 to 214 °C with a broadening of the peak. This behavior is interpreted as the mixing of a small amount of oligomer with the monomer crystal.

After 40-min irradiation, the DSP curve shows broad endothermic peaks around 300–350 °C. These peaks become larger (50 min) and are divided into two peaks after more than 70 min of irradiation which correspond to the two peaks of the DTA curve at 321 and 339 °C in Fig. 14. The smaller peak at 323 °C grows strikingly during the later stage of irradiation. The X-ray diffraction patterns of the samples irradiated for more than 50 min are almost the same as those of high molecular weight polymer crystals except that the higher the molecular weigth, the sharper are all the peaks.

Of further interest is the thermal depolymerization behavior of these polymers in the crystalline state. This behavior has been investigated by observing the continuous changes of the X-ray diffraction pattern and of the TG-DSC diagram during the heat treatment of poly-DSP[10, 65]. Changes of X-ray diffraction patterns on thermal treatment of as-polymerized poly-DSP crystals and photopolymerization of DSP crystals are shown in Fig. 16.

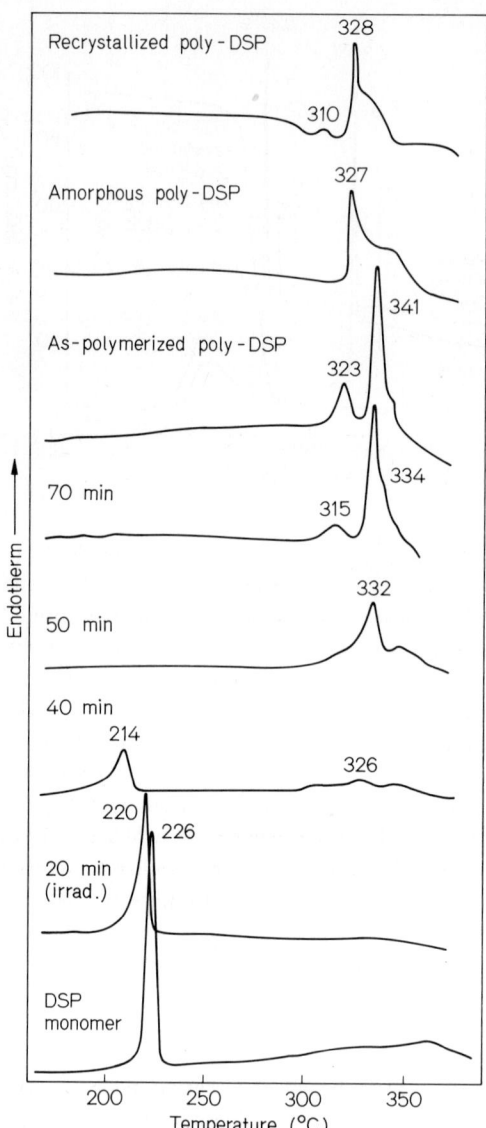

Fig. 15. DSC curves of recrystallized, amorphous, and as-polymerized poly-DSP's and the curves obtained in the course of the photopolymerization of DSP crystals. Light source; 500 w xenon lamp. (source: Ref. 10)

In Fig. 16 the heat-treated sample(1) is obtained by heating poly-DSP crystals (a) up to 330 °C at a scanning speed of 15 °C/min. Then, the sample is cooled immediately to room temperature. The intrinsic viscosity of the original as-polymerized poly-DSP (2.1–2.9) is reduced (0.55–0.59) in sample (1). An X-ray pattern of sample (1) shows slight but definite differences when compared to that of the original as-polymerized polymer and, in addition, the pattern agrees exactly with that ot the medium-sized polymer crystals (c), which are obtained by photopolymerization of DSP crystals upon irradiation with a xenon lamp for 50 min. DSC curves of sample (1) and polymer crystals (c) are also very similar to each other. From these results, it is concluded that the high

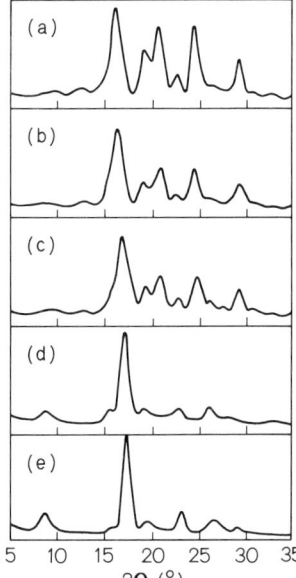

Fig. 16a–e. Changes of X-ray diffraction patterns upon heat treatment of as-polymerized poly-DSP crystals and photopolymerization of DSP crystals. (**a**) As-polymerized poly-DSP, (**b**) heat-treated sample (1), (**c**) poly-DSP obtained after 50-min irradiation of the monomer with a xenon lamp, (**d**) heat-treated sample (2), and (**e**) DSP oligomer obtained by selective monomer excitation. (source: Ref. 65)

molecular weight "as-polymerized" poly-DSP crystal has no definite melting point but depolymerizes to a relatively low molecular weight polymer crystal upon heat treatment.

The thermal depolymerization behavior in the crystalline state has been observed of most of polymers prepared by four-center photopolymerization, e.g. poly-p-PDA Me[31]. The X-ray diffraction patterns (d) and (e) in Fig. 16 will be discussed in Sect. VI.a.

Since two endothermic peaks, exemplified by the poly-DSP crystal in Fig. 14[64] are generally observed for the as-prepared polymer crystals and since a weaker peak appears at the lower temperature which grows at the later stage of photopolymerization as shown in Fig. 15[10], these two peaks are attributed to thermal depolymerization in the crystalline state and the processes involving degradation and/or crystal melting, respectively.

Based on the results of crystalline-state depolymerization, a reversible topochemical process, which is a monomer crystal lattice-controlled photopolymerization and a polymer crystal lattice-controlled thermal depolymerization, is established[65].

The process involving conversion of oligomer crystal to monomer crystal is not topochemical because at the oligomer stage the crystal melting point becomes lower than the thermal cleavage temperature of cyclobutane. From the comparison of the thermal depolymerization behavior by means of UV spectrometry and X-ray diffraction analysis, amorphous poly-DSP is much less stable to heat than as-polymerized poly-DSP crystals, presumably because the as-polymerized polymer crystal lattice protects, to a certain extent, the cyclobutane ring from thermal cleavage by restricting local movements of the polymer chain[65].

Enthalpy changes are derived from the peak areas of DSC curves taking the melting heat of DSP (12.0 kcal/mol) as a standard[66]. From the relative enthalpy levels of molten and crystalline DSP, and amorphous and crystalline poly-DSP, the topochemical photopolymerization of DSP is found to be an endothermal reaction with an enthalpy increase of about 3.7 kcal/mol.

It is very likely that during the polymerization under the strict control of the monomer crystal lattice, a strain energy caused by molecular movements accumulates on the reacting crystals resulting in the endothermal formation of polymer crystals. The enthalpy difference between the crystalline and amorphous polymers (3.1 kcal/mol) corresponds to the heat of crystallization of as-polymerized poly-DSP crystals.

VI. Polymer Properties

All the as-polymerized polymer crystals prepared by four-center photopolymerization are powdery white or slightly yellow substances and are stable in the atmosphere.

Studies on polymer properties have not been fully made for all the polymers, although poly-DSP has been investigated detail. However, some polymer properties are readily explained in terms of a polymer structure containing alternating 1,3-trans cyclobutane and 1,4-arylene repeating units in the main chain. Several unusual chemical and physical properties have been found. These concern the crystalline-state thermal depolymerization, the irreversible crystal structure of as-polymerized polymer, the thermal stability of the polymers, which is seriously affected by morphology and molecular weight, etc. These properties, which are presumably imparted to the polymer in a step-wise manner during the unusual process of crystal-lattice control, are charactristic of polymers prepared by four-center photopolymerization in the crystalline state.

a. Chemical Properties

All the polymers are highly resistant to alkaline, but not to oxidative conditions although only fragmental data on these properties are available. Poly-p-PDA Ph is barely hydrolyzed when dispersed in a 1.7 N aqueous sodium hydroxide solution and stirred at 70–80 °C for 24 h. Poly-p-PDA alkyl esters do not undergo alkaline hydrolysis at all under the same conditions[15]. Such high resistance of poly carboxylic esters toward hydrolysis may be attributed to the steric hindrance around the ester group branching site at the cyclobutane ring in the polymer chain. On the other hand, the cyano group in poly-p-CPA nPr is easily hydrolyzed by pouring a solution of the polymer in concentrated sulfuric acid into a large amount of water[67].

Most of the polymer crystals are thermally depolymerized to monomer and then degraded to undefinable materials[64, 68]. For example, an exothermal oxidation effect has been observed above 200 °C in the DTA of poly-DSP[64]. In aqueous solution, fine crystals of poly-DSP are readily oxidized with potassium parmanganate at 70 °C for 10 h, and 2,5-pyrazinedicarboxylic acid is identified as one of the reaction products containing a certain amount of benzoic acid[13].

The as-polymerized polymer crystal is fairly stable on photoirradiation in the air and faintly colored and becomes partially insoluble when photoirradiated for a long time, indicating cross-linking due to photooxidation.

In contrast to the relative stability of as-polymerized poly-DSP crystals in the air, amorphous poly-DSP is rapidly photooxidized in the air upon exposure to sunlight and produces nitrile derivatives together with products containing carbonyl and hydroxy groups, suggesting degradation of the pyrazine ring[69]. When amorphous poly-P 2 VB is photoirradiated in the air, cyano groups are produced at a slow rate. Furthermore, amorphous poly-p-PDA Et film is as stable as polystyrene under the same photooxidative condition. These stability differences suggest that the photooxidative behavior in these polymers cannot be attributed to tertiary hydrogens attached to the cyclobutane ring but to an unknown factor.

Most of the polymers are easily depolymerized photochemically and thermally in solution to the corresponding monomers, as is expected from the ring cleavage reaction of a number of cyclobutane derivatives yielding two olefins. For example, poly-DSP in solution is depolymerized to DSP nearly quantitatively upon photoirradiation for a relatively short period[36] or by heating at above 200 °C[70].

However, there are a few interesting exceptions. High molecular weight polymers from p-PDA and p-CPA derivatives are photodepolymerized to the oligomers in solution, but the oligomers are extraordinarily stable and scarcely converted to the monomer even after prolonged photoirradiation[34, 68]. Such photostable behavior has also been observed in the model reaction of α-truxillic acid ester (the 1,3-dimer of ethyl cinnamate, the model compound of a repeating unit of poly-p-PDA Et). α-Truxillic acid ester is readily photocleaved ethyl cinnamate whereas 1,3-diethoxycarbonyl-2,4-bis[p-(2-ethoxycarbonylvinyl)phenyl]cyclobutane (the 1,3-dimer of p-PDA Et, the model compound of p-PDA Et oligomer) is not converted into p-PDA Et at all[68].

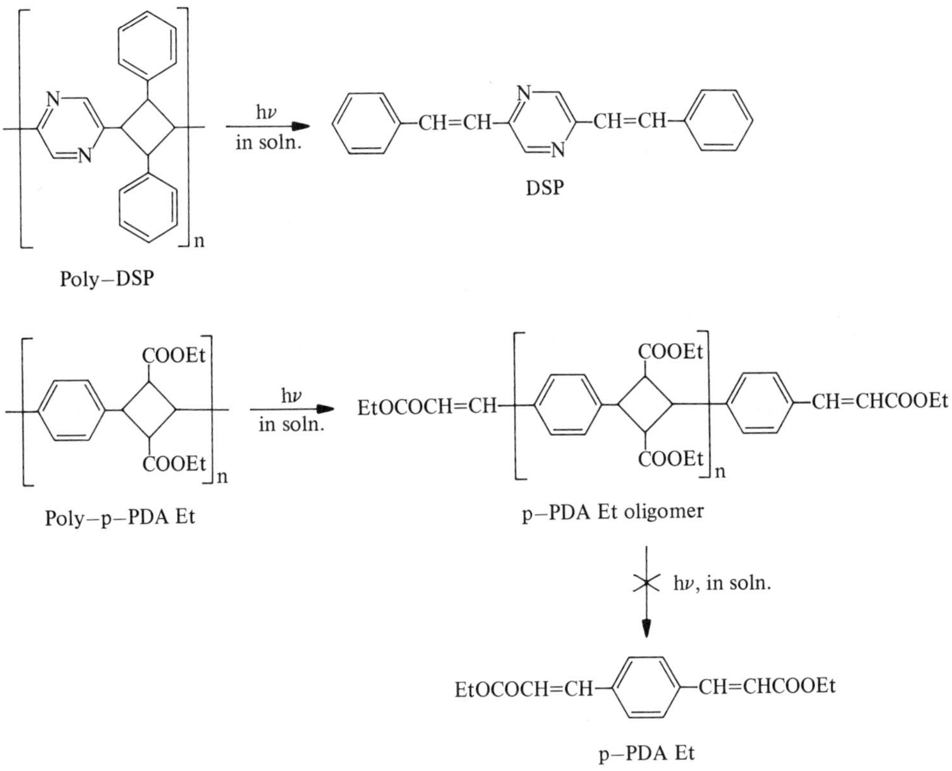

The excitation energy of the olefinic groups in these photostable compounds is probably directed not toward cyclobutane cleavage but largely toward cis-trans isomerization.

A prominent effect of wavelength on the photodepolymerization of oligo-p-CPA nPr in solution has been observed. An appreciable amount of monomer is produced when the oligomer solution is irradiated with monochromatic light of 224 nm whereas the oligomer is stable when irradiated at 304 nm[34].

It is of great interest that when the heat-treated sample (1) of poly-DSP in Fig. 16 is heated to 330 °C (with the same scanning speed), the X-ray diffraction pattern of the resulting sample (2) changes into a pattern which is exactly the same as that of the as-polymerized oligomer crystals prepared by selective monomer excitation (see Sect. III.b.).

It should be reminded that when DSP crystals are irradiated with a xenon or high-pressure Hg lamp without filter, the X-ray pattern never passes through the same pattern as that of the oligomer prepared by selective monomer excitation at the intermediate stage of photopolymerization[20]. In addition, the as-prepared oligomer crystals have space groups with identical those of the monomer and polymer crystals[26]. Such oligomer crystals are obtained by thermal depolymerization only when the as-polymerized poly-DSP crystal is the starting polymer.

Good coincidence of the X-ray patterns of these two kinds of oligomers suggests that the thermal chain scission in as-polymerized poly-DSP crystals does not proceed randomly at the position of cyclobutane ring in the molecular chain but is somewhat favored on the position of cyclobutane ring in the middle of the chain[65]. Thus, the oligomer crystal is preferentially accumulated in thermal depolymerization under the polymer crystal-lattice control. This abnormal chain scission is most plausibly explained when the

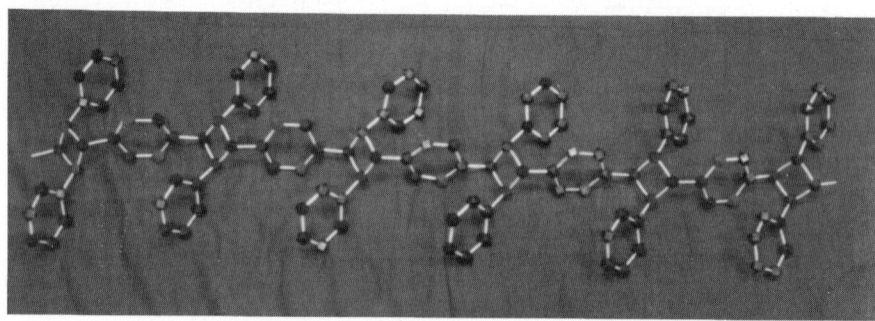

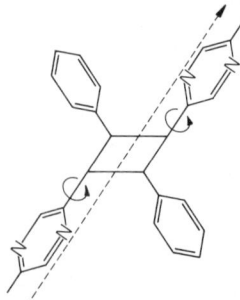

Fig. 17. Molecular model of poly-DSP. (source: Ref. 71)

rigid rod-shaped configuration of the polymer is taken into account. The simulated behavior is seen in the familiar phenomenon of a long slender stick being easily broken at the central part whereas a slender yet shorter stick is not broken by the same force.

All the polymers studies so far contain alternating 1,3-trans cyclobutane and 1,4-arylene units in the main chain. As a matter of course, these polymers are stiff rods even when the σ-bonds between the cyclobutane and aromatic rings in the chain rotate freely, as is obvious in poly-DSP (Fig. 17).

For the purpose of visualizing the above phenomenon of rigid rod-shaped sticks in terms of molecular dimensions the thermal stabilities of high and low molecular weight poly-DSP and poly-p-PDA Me have been compared both in the crystalline state and in solution[31, 70, 71]. When as-polymerized poly-DSP, for which the reduced viscosities are 3.0 and 8.4 dl/g are kept at 290 °C for a long period, the longer polymer chain of 8.4 dl/g is broken down to a polymer having a reduced viscosity of 4.0 dl/g. On the other hand, the shorter polymer of the reduced viscosity of 3.0 dl/g stays at the original value after 10 h[70]. In addition, no appreciable changes are seen in either the X-ray pattern or DSC scans of the same polymer. The decreased viscosity values are almost completely saturated after 10 h at various definite temperatures. The same temperature effect on molecular length has also been confirmed by thermal depolymerization of these polymers in solution although the upper limit of the average chain length in solution is much shorter than that in the crystalline state at the same temperature.

Changes of the molecular weight distribution curve of as-polymerized poly-DSP crystals by heat treatment at 265 °C support the explanation of the behavior by means of the rigid rod-shaped molecular model. This implies that by thermal treatment the distribution curve becomes narrower than the original curve with a decrese of the intrinsic viscosity from 5.1 to 3.9[72].

From these results the thermal stability which depends on molecular weight is conclusive of rigid rod-shaped molecules as illustrated by poly-DSP in Fig. 17. Furthermore, the same explanation can be applied to both the results in the crystalline state and in solution; shearing forces should always show the greatest stress at the middle of the rigid rod-like molecular chain.

From the extrapolation of the thermal stability of the polymer in solution, it is obvious that the true solution viscosity values for extremely high molecular weights of as-polymerized polymer crystals cannot be obtained since actual thermal depolymerization must occur during the solubilization procedure. In other words, there is a limit of chain length where the rod-shaped molecule can exist as a stable species under given conditions. Also, from the extrapolation of the thermal stability of poly-DSP crystals it is presumed that the polymer chain cannot grow endlessly in topochemical photopolymerization even in perfect crystals with no defects, but chain growth is in equilibrium with thermal depolymerization at a definite chain length under given conditions. This interpretation leads to the idea that such rigid rod-like molecules with extremely high molecular weights depolymerize thermally in the crystalline state even below room temperature.

In contrast to the rigid rod-shaped polymer, the polymer with a cyclobutane ring incorporated into the flexible main chain shows rather excellent thermal stability[73, 74]. Poly-(hexamethylene-α-truxillamide) is smoothly prepared by thermal polycondensation at ca. 200 °C or higher and is much more heat stable than the polymer containing cyclobutane rings in a rigid chain. The difference in the thermal stability between these two kinds of polymers, rigid or flexible, leads to the conclusion that the thermal stability of the polymers prepared by four-center photopolymerization is inherently depressed not only by the tendency of cyclobutane to undergo thermal eleavage but also by the rigid rod-like polymer shape.

This is a novel type of polymer effect due to a polymer chain shape where chemical bond stability is impaired by superimposed random shearing forces at some definite site of the molecule. A similar effect has been reported by Oster upon sonic treatment of tabacco mosaic virus where the polymer aggregates are dissociated[75].

As already described, amorphous poly-DSP is readily photooxidized upon exposure to the sunlight in the air[69]; however, the same sample is thermally depolymerized to the monomer in an oxygen-free atmosphere more rapidly than the crystalline polymer[65].

b. Physical Properties

As-polymerized polymers are highly crystalline and insoluble in most conventional organic solvents except strong acids such as concentrated sulfuric acid, dichloroacetic acid and trifluoroacetic acid. On the other hand, amorphous polymers are soluble in certain ordinary organic solvents. For example, amorphous poly-DSP, which is regenerated from solution, is soluble in m-cresol and o-chlorophenol at room temperature[76]. From this observation, the limited solubility of these as-polymerized polymers is explained not as a part of the repeating chemical structure but attributed to the extremely high crystallinity of as-polymerized crystals.

Crystallizatin of the amorphous polymer scarcely takes place upon thermal treatment but proceeds further in the presence of sufficient solvent thus enabling the polymer molecules to move and crystallize. The maximum degree of recrystallization has been achieved by keeping the amorphous film of poly-DSP in o-chlorophenol-ethanol (2:1 volume ratio) at 40 °C for a week[10]. The degree of crystallization of the film is about 15% according to Herman's method. It is noteworthy that of all the polymers examined so far the crystal stucture of regenerated polymers is different from that of as-polymerized polymers as X-ray diffraction patterns reveal[76,77]. Such an irreversible transition of crystal stuctures implies a change from a quasi-stable state in as-polymerized polymer to a stable state in recrystallized polymer, and originates in the unusual thermodynamic course of the four-center photopolymerization.

Although no prominent differences have been found between the DSC curves of recrystallized and amorphous polymers, a slight endotherm, which may correspond to the crystal melting point or thermal depolymerization of the recrystallized poly-DSP, is seen at 310 °C in the former curve (Fig. 15)[10].

Several physical properties of amorphous poly-DSP films have been measured[64]. For the preparation of amorphous films, a trifluoroacetic acid solution containing 10% poly-DSP is cast on a glass plate and dried in vacuo. The resulting film is extracted with triethylamine to remove the last traces of the acid and then extracted with ethanol.

The E' value determined by dynamic viscoelasticity measurements is 2×10^{11} dyne/cm^3 at room temperature. It decreases abruptly in the temperature range 140–150 °C, but the net decrease of E' within this temperature range is relatively small.

The electical properties of amorphous poly-DSP are characterized by a small temperature dependence of the dielectric constant measured between room temperature and 100 °C. The dielectric loss tangent is small and, in addition, the dc conductivity is extremely low.

Rotations of phenyl branches in the polymer seem to take place above $-30\,°C$, and glass transition occurs at about $150\,°C$. However, a certain depolymerization inevitably happens at the temperature above $100\,°C$.

Preliminary studies on the hydrodynamic behavior of poly-DSP and poly-p-PDA Et have been carried out. In contrast to the expectations based on rigid-chain stucture, non-Newtonian flow behavior is not observed for dilute solution viscosities of poly-DSP and poly-PDA Et in o-chlorophenol[76, 77]. However, a shear rate dependence of flow birefringence and a high value of the second virial coefficients have been found for a chloroform solution of poly-PDA Et, suggesting that the polymer molecule has a rod-like shape in dilute solution. From logarithmic plots of intrinsic viscosity and number average molecular weight, the following correlations have been determined for poly-DSP and poly-p-PDA Et in different solvents:

For poly-DSP[76]:

$[\eta] = 1.8 \times 10^{-2} \overline{M}_n^{0.54}$ (trifluoroacetic acid, $30\,°C$)
$[\eta] = 2.15 \times 10^{-2} \overline{M}_n^{0.52}$ (dichloroacetic acid, $30\,°C$)
$[\eta] = 25.6 \times 10^{-2} \overline{M}_n^{0.22}$ (o-chlorophenol, $30\,°C$)
$[\eta] = 21.8 \times 10^{-2} \overline{M}_n^{0.23}$ (m-cresol, $30\,°C$)

For poly-PDA Et[77]:

$[\eta] = 1.45 \times 10^{-3} \overline{M}_n^{0.68}$ (dichloroacetic acid, $30\,°C$)
$[\eta] = 3.23 \times 10^{-3} \overline{M}_n^{0.60}$ (o-chlorophenol, $30\,°C$)
$[\eta] = 3.52 \times 10^{-4} \overline{M}_n^{0.82}$ (chloroform, $30\,°C$)

In order to obtain highly oriented polymer crystals of rigid rod-shaped chain molecules, crystallization under shear stress and flow-induced crystallization have been attempted on concentrated sulfuric acid solutions of poly-DSP, poly-p-PDA Me and poly-p-CPA nPr[78]. During the crystallization process the branched cyano groups in poly-p-CPA nPr are partially hydrolyzed. Fibrillized polymer aggregates with a highly polarized molecular alignment have been observed on regenerated polymers under the polarizing microscope. The well-known liquid crystal behavior is seen on the same polymer when a concentrated solution is placed between highly sheared glass plates.

Preliminary tests using wet-spinning technique have been reported on poly-p-CPA nPr by applying the same combination of solvent and coagulant in the crystallization process[79]. No property of the filament which could be utilized in practice has been found so far although an extraordinarily high tenacity is anticipated from the liquid crystal intermediate, due to the rigid-chain structure.

Attempts have been made to apply new dry-process recording material to the polymerization of m-PDA crystals by using the continuous change of the refractive indices of the reacting crystals by Hattori et al[80]. Any practical use of the typical four-center photopolymerization, however, has not been reported so far.

VII. Characteristic Features of Four-Center Photopolymerization

The empirical rule[16] of the extention of dimerization to polymerization in the crystalline state is useful in the search for new photoreactive crystals in the future. However, a quantitative correlation between photoreactivity and molecular alignment in the crystal has not been established. Quite recently, Nakanishi has proposed a quantitative correlation between photoreactivity and intermolecular plane-to-plane distance in the crystal[81].

According to all the results of the studies on four-center photopolymerization, the whole reaction scheme of the DSP~poly-DSP system is given below showing the typical behavior of photopolymerizable conjugated diolefinic compounds[10].

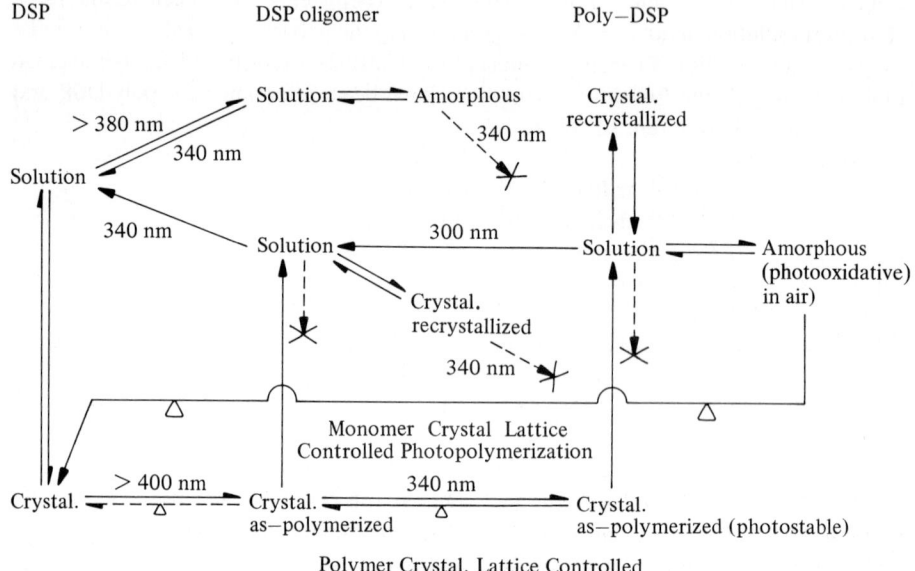

The reaction paths in solution are interpreted mostly by well-known reversible reactions between two olefins and a cyclobutane ring. On the other hand, in the crystalline state, a topotactic force makes these paths very unusual. Thus, neither recrystallized oligomer nor any other oligomer grow further by photoexcitation in solution because terminal olefinic bonds in such oligomers are not facing each other intermolecularly. In other words, there is no "topotactic assistance" in these oligomers[33]. In the polymerization, the reacting crystal lattice prevents thermal deterioration of the molecule by restricting thermal movements and as a consequence favors polymer chain growth. This is the reason why extended rigid long chains are produced only in the crystalline state, though chain growth is in equilibrium with thermal depolymerization.

A strain energy in the polymer crystallites caused by molecular displacement during polymerization is observed. In some cases, this is due to the fact that the polymer crystallites show interference color patterns under a polarizing microscope[20]. Such an accumulated strain energy is characterized as a distinctive feature of polymer property. This type of strain energy may play an important role in the crystal growth of natural polymers.

The relation between thermal behavior and molecular weight of the polymers involving thermal stability, which depends on molecular weight, is readily seen from Fig. 18.

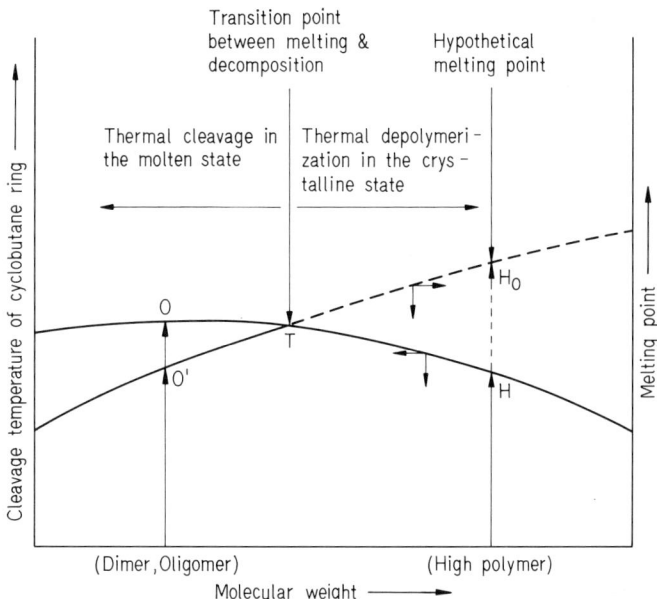

Fig. 18. Relation between thermal behavior and molecular weight of cyclobutane polymers with rigid rod-like structure. (source: Ref. 72)

Due to chain growth, the intermolecular forces of the extended polymer chain crystals would increase monotonously whereas the cyclobutane rings at the central part of the chain become thermally less stable with increasing molecular weight[10].

The cleavage temperature of cyclobutane in Fig. 18 is the temperature at the weakest point of the polymer chain which is at the central part. The melting point of the dotted line is hypothetical, because in this region an equilibrium between crystalline and molten states does not exist. On heating the polymer crystal in the region of hypothetical melting, the crystalline-state depolymerization proceeds along the rigid line from H to the transition point T. At the molecular weight below T, the crystal melting point becomes lower than the cleavage temperature of the cyclobutane ring, as indicated by points 0 and 0'.

The inversion from crystal melting to depolymerization may be visualized only on the polymers prepared by the lattice controlling polymerization, because propagation in such a process would be beyond the limit of the polymer chain growth under the reaction conditions including ordinary thermal diffusion. In other words, the molecular arrangement favorable for the polymerization stimulates excessive chain growth, leading to the formation of a long rigid polymer chain in the crystalline state.

Topotactic forces affect the ordinary reaction in some other ways. There is the case of a small molecule participating in crystalline-state photodimerization (see Sect. III.a.), e.g. small molecules such as water or n-hexane in the atmosphere are involved in the reaction[29, 30]. These small molecules, sometimes even the moisture in the air, not only greatly affect the rate and conversion of the reaction but are also included in the crystal of the photoproduct as a crystal molecule at the end of the reaction. In such dimer crystals, a small molecule may be useful for filling up the cave in the crystal, thus stabilizing the crystal skeleton. Although the role of the reacting crystal lattice has been demonstrated by several experimental results, further work is required to clarify the ultimate function of the crystals in this type of topochemical process.

Deterioration of the properties of amorphous poly-DSP and poly-P 2 VB on exposure to sunlight in the air is another striking abnormal behavior which is observed neither in the corresponding as-polymerized polymer crystals nor in the polymer in solution[69].

Considering these charactristics from an overall point of view, it is possible that in the four-center photopolymerization, both the polymerization process and the polymer properties may provide suitable patterns for naturally occurring polymers.

Acknowledgement. I am grateful to Dr. H. Nakanishi of the Research Institute for Polymer and Textiles (RIPT) for his assistance in the preparation of this article.

The major part of the research described in this article was undertaken at the RIPT, formerly at Yokohama and now at Tsukuba-gun Ibaragi-ken, Japan.

VIII. References

1. Liebermann, C.: Chem. Ber. 22, 124, 782 (1889)
2. Cohen, M. D., Schmidt, G. M. J.: J. Chem. Soc. *1964; 1969;* for a summary of the results see G. M. J. Schmidt: Pure Applied Chem. 27, 647 (1971)
3. Koelsch, C. F., Gumprecht, W. H.: J. Org. Chem. 23, 1603 (1958)
4. Suzuki, F., Suzuki Y., Hasegawa, M.: Bull. Tex. Res. Inst. 72, 11 (1965) (Japanese)
5. Hasegawa, M., Suzuki, Y.: J. Polym. Sci. B 5, 813 (1967)
6. Holm, M. J., Zienty, F.: US Pat. No. 3 312 688 CA 67, 12 151, 1967
7. Holm, M. J. Zienty, F.: J. Polym. Sci. A 1, 10, 1311 (1972)
8. For a summary of the results see M. Hasegawa et al.: Progr. Polym. Sci. Japan 5, 143 (1973)
9. Addadi, L., Lahav, M.: J. Am. Chem. Soc. 100, 2838 (1978); ibid. 101, 2152 (1979)
10. Hasegawa, M., Nakanishi, H., Yurugi, T.: J. Polym. Sci., Polym. Chem. Ed. 16, 2113 (1978)
11. Nakanishi, H. et al.: Proc. R. Soc. London A 369, 307 (1980)
12. Franke, R.: Chem. Ber. 38, 3727 (1905)
13. Hasegawa, M. et al.: J. Polym. Sci. A 1, 7 743 (1969)
14. Williams, J. L. R.: J. Org. Chem. 25, 1839 (1960)
15. Suzuki, F. et al.: J. Polym. Sci. A 1, 7, 2319 (1969)
16. Nakanishi, F., Hasegawa, M.: J. Polym. Sci. A 1, 8, 2151 (1970)
17. Nakanishi, F. et al.: J. Chem. Soc., Japan 1981, 412 (Japanese)
18. Meyer, W.: thesis, Univ. of Freiburg, Freiburg, W. Germany, 1977
19. Nakanishi, H. et al.: J. Polym. Sci. A 1, 11, 2501 (1973)
20. Nakanishi, H. et al.: ibid. A 1, 7, 753 (1969)
21. Miura, M., Kitami, T., Nagakubo, K.: Polym. Lett. 6, 463 (1968)
22. Nakanishi, F., Nakanishi, H., Hasegawa, M.: J. Polym. Sci., Polym. Chem. Ed. 13, 2499 (1975)
23. Watanabe, S. et al.: Unpublished data
24. Baker, W., Howes, C. S.ß J. Chem. Soc. *1953*, 119
25. Riiber, C. N.: Chem. Ber. 46, 335 (1913)
26. Nakanishi, H., Hasegawa, M., Sasada, Y.: J. Polym. Sci. A 2, 10, 1537 (1972)
27. Stobbe, H., Hensel, A.: Chem. Ber. 59, 2260 (1926)
28. Gerasimov, G. N. et al.: Dokl. Akad. Nau. SSSR 216, 1051 (1974)
29. Nakanishi, F. et al.: Bull. Chem. Soc., Japan 49, 3096 (1976)
30. Nakanishi, F., Yamada, Y., Nakanishi, H.: J. Chem. Soc., Chem. Comm. *1977*, 247
31. Hasegawa, M. et al.: J. Polym. Sci., Polym. Chem. Ed., in press

32. Tamaki, T., Suzuki, Y., Hasegawa, M.: Bull. Chem. Soc. *45*, 1988 (1972)
33. Hasegawa, M., Suzuki, Y., Tamaki, T.: Bull. Chem. Soc. *43*, 3020 (1970)
34. Suzuki, Y., Hasegawa, M., Kita, N.: J. Polym. Sci. A 1, *10*, 2473 (1972)
35. Higuchi, J., Ito, T.: private communication
36. Suzuki, Y., Tamaki, T., Hasegawa, M.: Bull. Chem. Soc., Japan, *47*, 210, (1974)
37. Nakanishi, F., Nakanishi, H., Hasegawa, M.: J. Chem. Soc.,Japan *1976*, 1575 (Japanese)
38. Sakamoto, M. et al.: Chem. Lett. *1981*, 99
39. Huy, S., thesis, Tokyo Inst. of Technol., Tokyo, Japan 1979; to be published
40. Iguchi, M., Nakanishi, H., Hasegawa, M.: J. Polym. Sci. A 1 *6*, 1055 (1968)
41. Nakanishi, H. Ueno, K., Sasada, Y.: Acta, Cryst. *B34*, 2209 (1978)
42. Sasada, Y. et al.: Bull. Chem. Soc., Japan, *44*, 1262 (1971)
43. Nakanishi, H. et al.: Chem. Lett. *1972*, 301
44. Nakanishi, H. Hasegawa, M., Sasada, Y.: J. Polym. Sci., Polym. Phys. Ed. *15*, 173 (1977)
45. Nakanishi, H., Sasada, Y.: Bull. Chem. Soc., Japan, *50*, 3182 (1977)
46. Nakanishi, H., Ueno, K., Sasada, Y.: Acta Cryst. *B32*, 1616 (1976)
47. Nakanishi, H., Ueno, K., Sasada, Y.: ibid. *B32*, 3352 (1976)
48. Nakanishi, H., Sasada, Y.: ibid. *B34*, 332 (1978)
49. Ueno, K. et al.: Ibid. *B34*, 2034 (1978)
50. Nakanishi, H., Ueno, K., Sasada, Y.: ibid. *B34*, 2036 (1978)
51. Nakanishi, H., Ueno, K., Sasada, Y.: J. Polym. Sci., Polym. Phys. Ed. *16*, 767 (1978)
52. Baughman, R. H.: J. Appl. Phys. *41*, 4579 (1971)
53. Nakanishi, H. et al.: Israel J. Chem. *18*, 261 (1979)
54. S. Huy et al.: unpublished data
55. Chatani, Y. et al.: J. Polym. Sci. A 2, *11*, 369 (1973)
56. Wegner, G.: Makromol. Chem.: *154*, 35 (1972)
57. Meyer, W., Lieser, G., Wegner, G.: Makromol. Chem. *178*, 631 (1977); J. Polym. Sci., polym. Phys. Ed. *16*, 1365 (1978)
58. Nakanishi, H., Sasada, Y., Hasegawa, M.: Polym. Lett. Ed. *17*, 459 (1979)
59. Jones, W.: J. Chem. Research *1978*, 142
60. Jones, W., Thomas, J. M.: Prog. in solid State Chem., 101 (1979)
61. Nakanishi, H. et al.: J. Chem. Soc., Japan *1977*, 1046 (Japanese)
62. Nakanishi, H., Nakano, N., Hasegawa, M.: J. Polym. Sci., Polym. Lett. *8*, 755 (1970)
63. Nakanishi, H., Jones, W., Thomas, J. M.: Chem. Phys. Lett. *71*, 44 (1980)
64. Kanetsuna, H. et al.: J. Polym. Sci A 2, *8*, 1027 (1970)
65. Hasegawa, M. et al.: Polym. Lett. *12*, 57 (1974)
66. Nakanishi, H., Hasegawa, M., Yurugi, T.: J. Polym. Sci., Polym. Chem. Ed. *14*, 2079 (1976)
67. Komatsu, K.: thesis, Tokyo Inst. Agricult. and Technol., Tokyo, Japan 1978; to be published
68. Nakanishi, F., Hasegawa, M., Tasai, T.: Polymer *16*, 218 (1975)
69. Sakuragi, M., Hasegawa, M., Nishigaki, M.: J. Polym. Sci., Polym. Chem. Ed. *14*, 521 (1976)
70. Hasegawa, M., Nakanishi, H., Yurugi, T.: Polym. Lett. *14*, 47 (1976)
71. Hasegawa, M., Nakanishi, H., Yurugi, T.: Chem. Lett. *1975*, 497
72. Hasegawa, M. et al.: to be published
73. Takahashi, H. et al.: J. Polym. Sci. A 1, *10*, 1399 (1972)
74. Nakanishi, F., Hasegawa, M., Takahashi, H.: Polymer *14*, 440 (1973)
75. Oster, G.: J. Gen Physiol. *31*, 89 (1947)
76. Fujishige, S., Hasegawa, M.: J. Polym. Sci. A 1, *7*, 2037 (1969)
77. Fujishige, S., Mochida, J., Tsuneo, T.: Bull. Res. Inst. Polym. and Tex. *107*, 1 (1975) (Japanese)
78. Yurugi, T. et al.: Prepring, Conf. Soc. Fiber Sci. and Technol, June 1978, Tokyo, Japan
79. Hasegawa, M.: to be published.
80. Mizuno, T., Tawata, M., Hattori, S.: J. Opt. Soc. Am. *67*, 1651 (1977)
81. Nakanishi, H.: thesis, Tokyo Inst. Technol., Tokyo, Japan 1978

Received March 10, 1981
T. Saegusa (editor)

Macrozwitterion Polymerization

David Samuel Johnston*

Chemical Laboratory, Trinity College, Dublin 2, Ireland

In the vast majority of ionic polymerizations reported in the literature the counterion is ionically bound to the polymer chain. When the counterion is covalently bound to the chain, the polymerization is termed macrozwitterionic. This review surveys what is known about this relatively unexplored area of polymer chemistry. A brief history of the topic is followed by a summary of the literature. The evidence presented in each report for the formation of macrozwitterions is critically assessed. Then an attempt is made to draw out features common to all monomer/initiator combinations which can thus be considered characteristic of macrozwitterion polymerization. These are contrasted to what would be thought typical of an ion pair polymerization. Macrozwitterionic polymerization offers a convenient route to macrocyclic ligands and, when polymeric initiators are used, a range of novel graft copolymers.

1	Introduction	53
2	**Literature Review**	55
2.1	Addition Polymerizations	55
2.1.1	Vinyl Monomers	55
2.1.2	Carbonyl Monomers	75
2.1.3	Strained Ring Monomers	78
2.2	Charge Cancellation Polymerization	86
2.2.1	Aryl Substituted Cyclic Sulfonium Zwitterions	86
2.2.2	Acetonitrile Oxide	88
2.2.3	Diazoalkanes	92
2.2.4	Spontaneous Alternating Copolymerizations	92
3	**Mechanism of Macrozwitterion Polymerization**	95
3.1	Covalent Initiators	96
3.2	Reactivity of Covalent Base – Monomer Adducts	97
3.3	The Influence of Chemically Inert Salts on the Polymerization Rate	99
3.4	The Role of Solvent in Charge Separation	100
3.5	Zwitterion Propagation	101

* Present address: Royal Free Hospital School of Medicine, University of London, Department of Biochemistry and Chemistry, 8 Hunter Street, London WC 1 N 1 BP, United Kingdom

4	**Conclusions**	102
4.1	Anionic Polymerization	102
4.2	Cationic Polymerization	102
4.3	Charge Cancellation Polymerization	104
5	**References**	104

1 Introduction

When ionic polymerizations are categorized it is usually according to the polarity of the active centre – anionic or cationic. There is, however, a more fundamental classification, based on the counter ion: is it or is it not covalently bound to the growing polymer chain?

$$X^{\oplus}(M)_n M^{\ominus} \qquad HO(M)_n M^{\ominus} B^{\oplus}$$
$$Y^{\ominus}(M)_n M^{\oplus} \qquad H(M)_n M^{\oplus} A^{\ominus}$$
$$A \qquad\qquad\qquad B$$

M = monomeric unit

The vast majority of ionic polymerizations studied fall into category B. However, there is now a substantial and growing literature about category A, for which the term "macrozwitterion" has been coined.

Beside charge location two factors distinguish macrozwitterion (from ion pair) polymerizations.

The initiator is a polarized molecule. Frequently ionization and charge separation take place during initiation. In the presence of ion transferring agents macromolecular salts are formed. As a corollary to this, polymerizations initiated by molecules, of which ion transferring agents are an integral part, cannot be classified as macrozwitterionic. Under these circumstances ions not covalently bound to the growing chains are present. Thus:

Initiation
$$X + M \rightleftharpoons {}^{\oplus}XM^{\ominus}$$
$$Y + M \rightleftharpoons {}^{\ominus}YM^{\oplus}$$

Propagation
$${}^{\oplus}XM_n M^{\ominus} + M \rightarrow {}^{\oplus}XM_{n+1}M^{\ominus}$$
$${}^{\ominus}YM_n M^{\oplus} + M \rightarrow {}^{\ominus}YM_{n+1}M^{\oplus}$$

Termination (by added ion transfer agents)
$${}^{\oplus}XM_{n+1}M^{\ominus} + HA \rightarrow (A^{\ominus})({}^{\oplus}XM_{n+1}MH)$$
$${}^{\ominus}YM_{n+1}M^{\oplus} + BOH \rightarrow (B^{\oplus})({}^{\ominus}YM_{n+1}MOH)$$

Zwitterions are commonly postulated as intermediates in chemical reactions. For example see Eq. (1)–(3) page 53.

The literature concerning such reactions can be helpful when attempts are made to elucidate the mechanism of initiation of macrozwitterion polymerizations. The influence on yield and rate of temperature, solvent, and the nature of the substituents on the atoms which formally carry charge, have been closely examined. It would be inappropriate to describe this work here, but the earliest notions about the formation of bipolar species from molecules is contained in it.

Before general acceptance of the co-catalysis mechanism for cationic polymerizations initiated by Friedel-Crafts halides, zwitterions were regarded as possible intermediates[4-6]. However, Horner seems to have been first to correctly identify a polymerization in which zwitterions are formed. Horner, Jurgeleit and Klupfel[7] studied the polymerization of acrylonitrile by triethylphosphine and reported their findings in 1955.

In a review of anionic polymerization in 1960[8], Swarc notes that bubbling isobutene through methyl cyanoacrylate yields a 1:1 copolymer. He suggests that a "highly polar

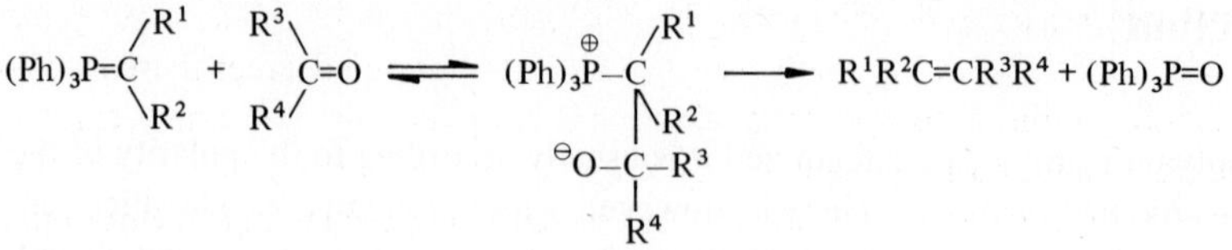

$$(Ph)_3P=C\begin{matrix}R^1\\R^2\end{matrix} + \begin{matrix}R^3\\R^4\end{matrix}C=O \rightleftharpoons (Ph)_3\overset{\oplus}{P}-\underset{\underset{R^4}{\overset{\ominus}{O-C-R^3}}}{\overset{R^1}{\underset{|}{C}}\begin{matrix}\\R^2\end{matrix}} \longrightarrow R^1R^2C=CR^3R^4 + (Ph)_3P=O \qquad (1)$$

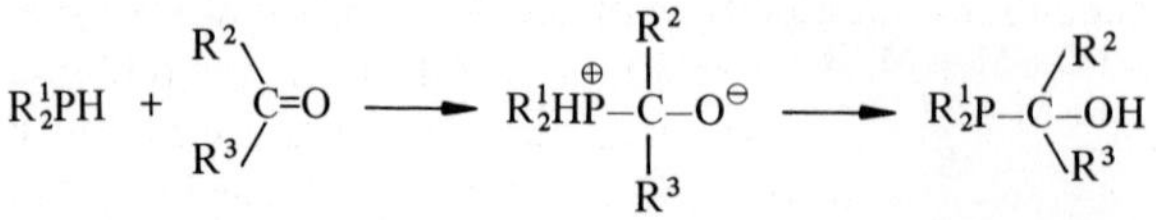

$$R_2^1PH + \begin{matrix}R^2\\R^3\end{matrix}C=O \longrightarrow R_2^1\overset{\oplus}{H}P-\underset{R^3}{\overset{R^2}{\underset{|}{C}}}-O^{\ominus} \longrightarrow R_2^1P-\underset{R^3}{\overset{R^2}{\underset{|}{C}}}-OH \qquad (2)$$

$$\begin{matrix}H_2C-CH_2\\|\quad\;\;|\\O-C=O\end{matrix} + HNR_2 \begin{cases} \longrightarrow \overset{\oplus}{R_2N}HCH_2CH_2CO_2^{\ominus} \\ \qquad\qquad\downarrow \\ \qquad R_2NCH_2CH_2COOH \\ \\ \longrightarrow \overset{\oplus}{R_2N}H-\overset{O}{\overset{\|}{C}}-CH_2CH_2-O^{\ominus} \\ \qquad\qquad\downarrow \\ \qquad R_2N-\overset{O}{\overset{\|}{C}}-CH_2CH_2OH \end{cases} \qquad (3)$$

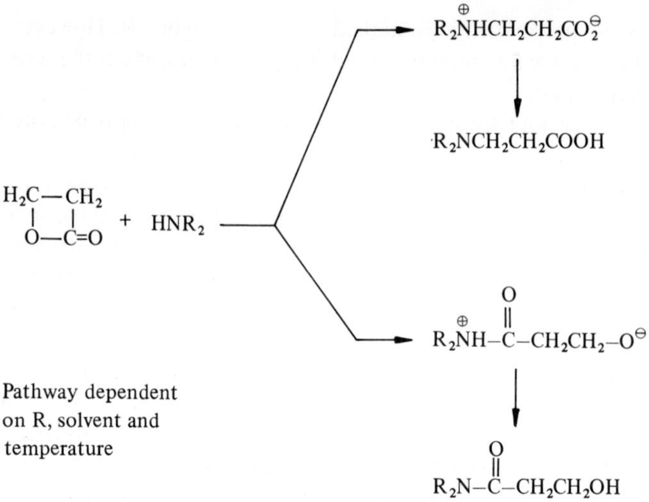

Pathway dependent on R, solvent and temperature

dimer is produced which polymerizes by virtue of its dipolar nature" and prophetically predicts that in such a polymerization a considerable amount of cyclic tetramer would form.

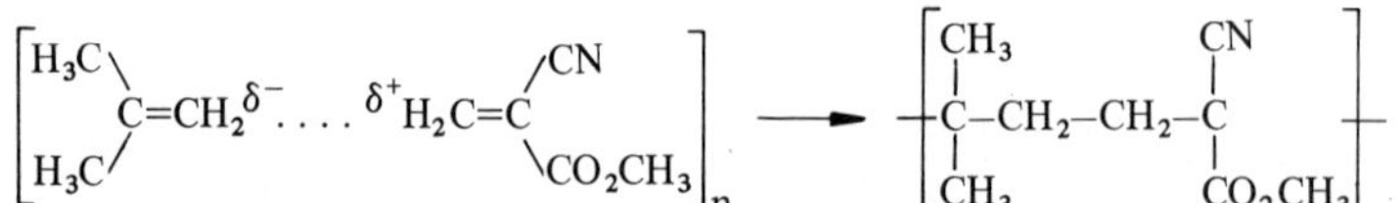

$$\left[\begin{matrix}H_3C\\H_3C\end{matrix}C=CH_2\delta^-\ldots\delta^+H_2C=C\begin{matrix}CN\\CO_2CH_3\end{matrix}\right]_n \longrightarrow \left[\begin{matrix}CH_3\\|\\-C-CH_2-CH_2-C-\\|\\CH_3\end{matrix}\begin{matrix}CN\\|\\\\|\\CO_2CH_3\end{matrix}\right]_n$$

Investigations directed specifically at examining polymerizations thought to proceed via macrozwitterions, begin in the early sixties. Monomers are unpolarized and thermodynamically stable. Such monomers generate reactive ions prone to transfer and termination processes. Chains partnered by independent counterions are introduced, and the study of monomer addition to macrozwitterions is difficult if not impossible. Consequently, the work is of very uneven quality, and later investigations have shown that some of the conclusions drawn are completely erroneous.

In the late sixties the emphasis shifts to cyclic monomers, primarily the lactones. An absence of transfer and termination processes enabled Jaacks[22] to conclusively demonstrate the existence of macrozwitterions and examine some peculiarities of their growth.

The last few years have seen the synthesis of zwitterions which polymerize by the charge cancellation mechanism foreshadowed by Swarc. To varying degrees these monomers do not exhibit a unique propagation reaction, charge cancellation can occur between monomers, chain and monomer, or chain and chain. However, there can be no doubt about the nature of the chains and one novel reaction predicted for zwitterions, macrocycle formation, has been shown to occur.

This review deals with polymerizations initiated by a molecule, in which at least one author believes macrozwitterions are formed. The majority are of the simple addition type and are subdivided into vinyl, carbonyl, and strained ring. The other types are all charge cancellations. Monomers are either preformed zwitterions, highly polarised molecules (i.e. CH_3CNO) or zwitterions are formed in situ by nucleophile/electrophile pairs.

2 Literature Review

2.1 Addition Polymerizations

2.1.1 Vinyl Monomers

Methacrylonitrile

The least electrophilic monomer in this category which is reputed to form macrozwitterions is methacrylonitrile.

Klippert and Ringsdorf[9, 10] tested the activity of a number of phosphoniomethylides as initiators of anionic polymerization. Their findings are summarized in Table 1.

Under the definition of a macrozwitterion polymerization given in the introduction, polymerizations initiated by salt complexed phosphoniomethylides are ruled out. Ions not covalently bound to the growing polymer chains will be present.

Table 1. Initiator activity of phosphoniomethylides (3 mole-%). Temperature: $-60\,°C$; Time = 120 min; solvent: toluene

Monomer	$(C_6H_5)_3\overset{\oplus}{P}-\overset{\ominus}{C}HCOCH_3$	$(C_6H_5)_3\overset{\oplus}{P}-\overset{\ominus}{C}HSi(CH_3)_3$	$(C_6H_5)_3\overset{\oplus}{P}-\overset{\ominus}{C}H_2$	$(C_6H_5)_3\overset{\oplus}{P}-\overset{\ominus}{C}H_2 \cdot LiBr$
Styrene	−	−	−	−
2-Vinyl-pyridine	−	−	−	−
Methyl-methacrylate	−	−	−	+
Acrylonitrile	−	+	+	+
Methacrylo-nitrile	−	+	+	+
Dialkyl methylene-malonate	+	+	+	+

Initiation of methacrylonitrile polymerization with triphenylphosphoniomethylide is formulated as follows:

$$(C_6H_5)_3\overset{\oplus}{P}-\overset{\ominus}{CH_2} + CH_2=\underset{CH_3}{\overset{CN}{C}} \longrightarrow (C_6H_5)_3\overset{\oplus}{P}-CH_2CH_2-\underset{CH_3}{\overset{CN}{\overset{|}{C}{}^\ominus}}$$

Two propagation steps are considered: insertion into the macrocycle (i) or into the linear aggregate (ii).

$$\overset{M}{\nearrow}\overset{\ominus}{M}\ R_3\overset{\oplus}{P}\quad (i)\quad ...M^\ominus-(R_3)P^\oplus...M^\ominus-(R_3)P^\oplus...\quad (ii)$$

The authors favour reaction (ii), because they believe that steric hindrance would make the formation of oligomeric rings from α-methylvinyl compounds very difficult.

Proof of the presence of macrozwitterions comes from end group analysis and mass spectroscopic measurements. Both triphenyl- and tricyclohexylphosphoniomethylides yield polymers with one phosphorus atom per chain. In the mass spectra of polymers initiated by triphenylphosphonio methylide (but not BuLi), fragments are found with high m/e-values characteristic of degradation products from triphenylphosphine and phosphine oxide.

The polymerization of this monomer by triethylphosphine has been extensively studied by Ranogayets, Kotchetov, Markevich, and Yenikolopyan[11–14].

The evidence they put forward for the existence of macrozwitterions is convincing. Elemental analysis of carefully purified polymers indicated one phosphorus atom per chain. NMR spectroscopy showed that the chemical shifts of ^{31}P in low molecular weight polymer and 2-cyanopropyltriethylphosphonium chloride as model compound were similar, 37.7 and 38.5 ppm, respectively. ^1H NMR spectra of the model compound with added potassium hydroxide and low molecular weight oligomers (OH^- counter ion) were virtually identical.

In the presence of water, conversion of monomer to polymer is accompanied by a parallel increase in solution conductance, the magnitude of the increase being proportional to the amount of water added. This is to be expected if zwitterions are formed.

$$\underset{H_2O}{\overset{\oplus}{\rule{2cm}{0.4pt}}\overset{\ominus}{}} \rightarrow \underset{OH^\ominus}{\overset{\oplus}{\rule{2cm}{0.4pt}}}H$$

The authors concentrate mainly on the kinetics of macrozwitterion growth. No simple order is apparent from plots of log (initial rate) versus log [Initial monomer].

At low active centre concentrations encountered, initiation is slow – termination rapid, ion pair ⇌ free ion equilibria are thought to be essentially intramolecular. As chain length increases, falling entropy will favour the free ion. Since free ion propagation is believed to be much faster than propagation on cyclic contact ion pairs, the authors suggest that k_p depends on the length of the zwitterion.

Macrozwitterion Polymerization

The monomolecular equilibrium constant for ion pairs ⇌ free ions, is defined as

$$\ln K_d = -\frac{\Delta F_e + \Delta F_c}{k_t}$$

ΔF_e and ΔF_c are, respectively, the electrostatic and chain free energy changes associated with the free ion → ion pair transition. An expression relating K to chain length is derived from its defining equation and polymer chain statistics

$$K_d^{(n)} = K_d^{(1)} n^{3/2} \; ; \; K_d^{(n)} = [R_n]/[r_n]$$
R_n: free ions, $DP = n$; r_n: ion pairs, $DP = n$

The kinetic scheme proposed for polymerization is:

Initiation	$C + M \to r_1$	$C = (C_2H_5)_3P$ $M = $ monomer
Propagation (Ion Pairs)	$r_1 + M \xrightarrow{k_\pm} r_2$	$r = $ ion pair
	$r_n + M \xrightarrow{k_\pm} r_{n+1}$	
	$r_n \underset{}{\overset{K_d^n}{\rightleftarrows}} R_n$	$R = $ free ion
Propagation (Free Ions)	$R_n + M \xrightarrow{k_-} R_{n+1}$	
Monomolecular Termination (Ion Pairs)	$r_n \xrightarrow{k_t} P_n$	$P = $ polymer
Monomolecular Termination (Free Ions)	$R_n \xrightarrow{k_t^1} P_n$	

Termination is thought to be monomolecular, because there is no conductance change in the absence of water.

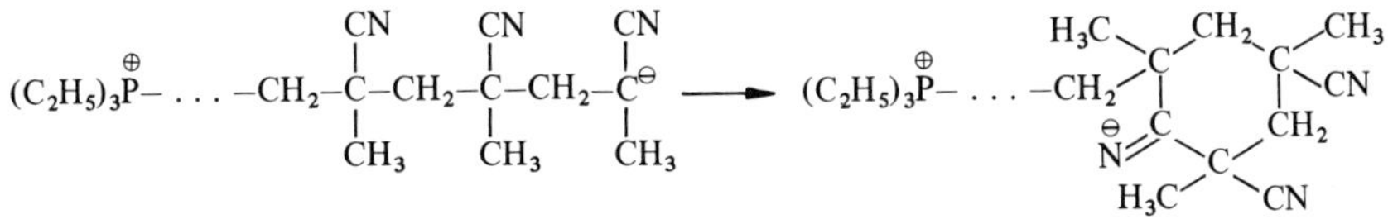

$$(C_2H_5)_3\overset{\oplus}{P}- \ldots -CH_2-\underset{CH_3}{\overset{CN}{\underset{|}{\overset{|}{C}}}}-CH_2-\underset{CH_3}{\overset{CN}{\underset{|}{\overset{|}{C}}}}-CH_2-\underset{CH_3}{\overset{CN}{\underset{|}{\overset{|}{\overset{\ominus}{C}}}}} \longrightarrow (C_2H_5)_3\overset{\oplus}{P}- \ldots -CH_2-\text{[cyclic structure]}$$

Thus:

$$-\frac{d[M]}{dt} = (k_\pm)[M]\sum[r_n] + (k_-)[M]\sum[R_n]$$

and using stationary state approximation

$$\frac{d([r_n] + [R_n])}{dt} = (k_\pm)[M][r_{n-1}] - (k_\pm)[M][r_n] + (k_-)[M][R_{n-1}]$$
$$- (k_-)[M][r_n] - (k_t)[r_n] - k_t^1[R_n] = 0$$

combining with $K_d^{(n)} = K_d^{(1)} n^{3/2}$, rearranging and simplifying gives

$$-\frac{d[M]}{dt_0} = \frac{k_i k_-}{k_t^1} [C_0][M_0^2] \exp\left[-\frac{2k_t}{(k_-)K_d^{(1)}[M_0]}\right]$$

Clearly, the order can rise above two. The above equation can be put in the following form

$$\log\left(\frac{d[M]}{dt_0}/[M_0]^2\right) = \log (k_i k_-/k_t^1)[C_0] - 1/2.3 \frac{2k_t}{K_d^{(1)} k_-} \frac{1}{[M_0]}$$

A plot of $\log \left(\frac{d[M]}{dt_0}/[M_0]^2 \right)$ versus $1/[M_0]$ was found to be linear.

The polymer molecular weight at fixed conversion was proportional to $[M]^{1.7}$. This value, greater than one, is believed to be linked with the self accelerated growth of polymer chains.

The theoretical treatment above is based on the assumption that ion pairing is intramolecular. Hence, it is only valid for low conversions of monomer to polymer. If low molecular weight polymer or a phosphonium salt were introduced the polymerization rate fell by two orders of magnitude. At higher polymer concentrations it seems that intermolecular contact ion pairs form. Added water also decreases the rate, perhaps partly because termination reduces the average molecular weight of the polymer at a given conversion.

In the absence of cyclic ion pairs the polymerization order with respect to monomer is two.

The kinetic equations so far derived permit determination of ratios of rate constants but not their individual values. The polymerization has an induction period which the authors attribute to a non-stationary active centre concentration. The following equation was derived to fit this phase of the polymerization

$$\text{I.P.} = \frac{1}{k_t^1} + \frac{2}{(k_-)K_d^{(1)}} \frac{1}{[M_0]}$$

and with it determination of individual rate constants is possible.

At 40 °C (in DMF):

$k_{\text{initiation}}$ = 5.6×10^{-4} l · mol^{-1} · min^{-1}
$k_{\text{prop(free ions)}}$ = 2.5×10^4 l · mol^{-1} · min^{-1}
$k_{\text{prop(ion pairs)}}$ = 2.0 l · mol^{-1} · min^{-1}
$k_{\text{term(free ions)}}$ = 4.0 min^{-1}

$k_{term(ion\ pairs)} = 0.84\ min^{-1}$
$K_d^{(1)} = 10^{-4}\ mol \cdot l^{-1}$
$K_3 = 0.07 \times 10^{-4}\ mol \cdot l^{-1}$

K_3 is the dissociation constant of linear ion pairs

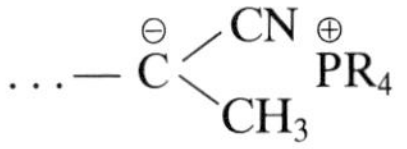

$$\ldots-\overset{\ominus}{C}\begin{matrix}CN \\ \ \\ CH_3\end{matrix}\overset{\oplus}{PR_4}$$

Rate constants for termination on water were determined by expanding the equation for the initial rate of polymerization:

$$\frac{d[M]}{dt_0} = \frac{k_i(k_-)[C_0][M_0]^2}{k_t^1 + k_{H_2O}[H_2O]} \exp^{-2(k_t + k_{H_2O}[H_2O])/(k_-)K_d^{(1)}[M_0]}$$

which can be put in the form

$$[C_0][M_0]^2 \Big/ \frac{d[M]}{dt_0} = \frac{k_t^1 + k_{H_2O}^1[H_2O]}{k_i k_-} \exp^{2(k_t + k_{H_2O}[H_2O])/(k_-)K_d^{(1)}[M_0]}$$

A plot of $[C_0][M_0]^2 \Big/ \frac{d[M]}{dt_0}$ versus H_2O was linear, suggesting negligible termination of ion pairs on water. $K_{H_2O} \rightarrow 0$

$k_{H_2O}^1 = 1.32 \times 10^4\ l \cdot mol^{-1} \cdot min^{-1}$

It will be noticed that on the subject of the existence of cyclic ion pairs Ringsdorf and Klippert and the Russian group are in disagreement.

Acrylonitrile

The literature contains descriptions of acrylonitrile polymerization initiated by three different phosphorus bases, triethyl phosphine, triphenylphosphine, and triethyl phosphite.

Ogawa et al.[15, 16] compared triethyl phosphite with triphenylphosphine. Polymerizations initiated by phosphite were faster and yielded undiscoloured polymers of higher molecular weight. It was preferred as initiator.

The more polar the solvent the faster the polymerization rate. In benzene the polymerization order with respect to initiator was one. The order with respect to monomer rose from two at $-5\,°C$ to above three at $+40\,°C$. The following three step kinetic scheme was proposed to account for this.

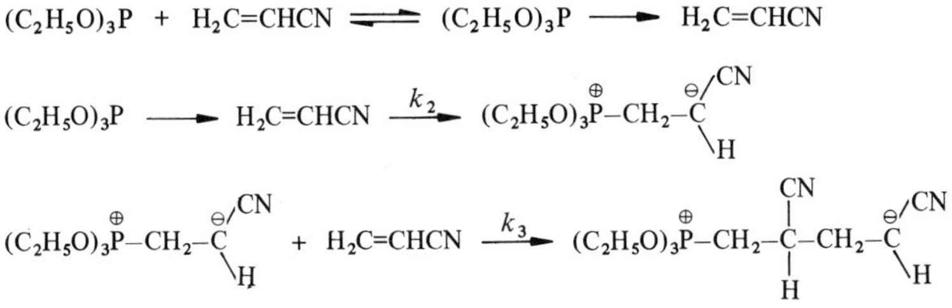

$$(C_2H_5O)_3P + H_2C=CHCN \rightleftharpoons (C_2H_5O)_3P \longrightarrow H_2C=CHCN \qquad (4)$$

$$(C_2H_5O)_3P \longrightarrow H_2C=CHCN \xrightarrow{k_2} (C_2H_5O)_3\overset{\oplus}{P}-CH_2-\overset{\ominus}{C}\begin{matrix}CN \\ \ \\ H\end{matrix} \qquad (5)$$

$$(C_2H_5O)_3\overset{\oplus}{P}-CH_2-\overset{\ominus}{C}\begin{matrix}CN \\ \ \\ H\end{matrix} + H_2C=CHCN \xrightarrow{k_3} (C_2H_5O)_3\overset{\oplus}{P}-CH_2-\overset{|}{\underset{|}{C}}\begin{matrix}CN \\ \ \\ H\end{matrix}-CH_2-\overset{\ominus}{C}\begin{matrix}CN \\ \ \\ H\end{matrix} \qquad (6)$$

At low temperatures reaction (5) is thought to be very slow. Assuming monomolecular termination and a steady state in active centre concentration at the start of polymerization

$$k_2 \text{ [C.T. complex]} = k_t [AN^\ominus]$$

$$K = \frac{\text{[C.T. complex]}}{[(C_2H_5O)_3P]_0[AN]_0} \quad \text{[C.T. complex]} = K[(C_2H_5O)_3P]_0[AN]_0$$

$$R_i = k_p \cdot [AN^-][AN]_0 \quad R_i = \text{initial rate of polymerization}$$

$$R_p = \frac{k_p k_2}{k_t} K[(C_2H_5O)_3P]_0[AN]_0^2$$

However, as the temperature rises the rate of reaction (5) is believed to increase more rapidly than reaction (6).
Eventually:

$$k_3[(C_2H_5O)_3\overset{\oplus}{P} \; CH_2\overset{\ominus}{C}{<}^{CN}_{H} \; [AN]_0 = k_t[AN^\ominus]$$

and

$$R_p = \frac{k_3 k_p}{k_t} K[AN]_0^3 [(C_2H_5O)_3P]_0$$

If the polymerization was carried out in dimethylformamide the order was two, regardless of the temperature. Presumably, k_3 is larger in the more polar solvent.

To explain the parallel fall in molecular weight and phosphorus content of the polymer with increasing temperature Ogawa and Quintana[17] have postulated transfer to monomer.

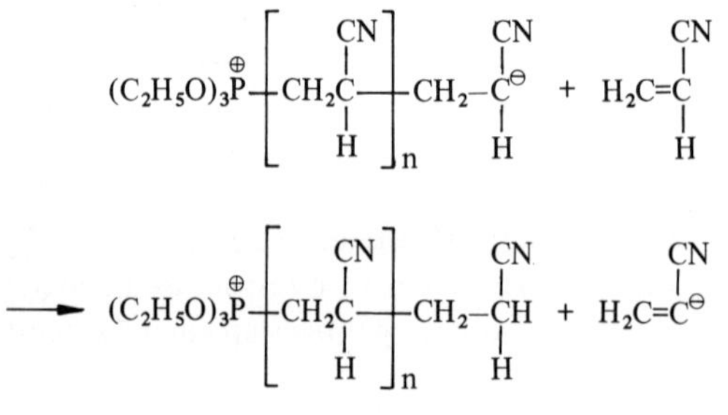

$$(C_2H_5O)_3\overset{\oplus}{P}{-}\!\!\left[CH_2\underset{H}{\overset{CN}{C}}\right]_n\!\!{-}CH_2{-}\overset{CN}{\underset{H}{C}}{}^\ominus + H_2C{=}\overset{CN}{\underset{H}{C}}$$

$$\longrightarrow (C_2H_5O)_3\overset{\oplus}{P}{-}\!\!\left[CH_2\underset{H}{\overset{CN}{C}}\right]_n\!\!{-}CH_2{-}\overset{CN}{\underset{H}{C}}H + H_2C{=}\overset{CN}{C}{}^\ominus$$

This reaction would have to have a higher activation energy than initiation or propagation. Only at low temperature, $(-10\ °C)$ when polymer formed is found to have one covalently bound phosphorus atom per chain, are all growing chains macrozwitterions.

Like Klippert and Ringsdorf, these authors suggest that propagation involves monomer insertion into a macrozwitterion aggregate.

More was learnt about the initiation mechanism from polymerization in the presence of added salt and the donor, triethylamine. Tetraethylammonium bromide (TEAB) and triethylamine increase the polymerization rate.

The salt caused a sixty fold increase when the solvent was 1,4-dioxane. Two explanations are put forward to account for the effect of TEAB. Firstly that monomer adds more rapidly to a quaternary ammonium ion-carbanion pair or secondly that by minimizing the interaction between the poles of the zwitterion, the bromide ion increases the rate constant, k_3.

There is some evidence to support the second suggestion. With acetonitrile as solvent the initial rate of polymerization at 40 °C is third order with respect to monomer. In the presence of salt the order falls to two. A similar change in order which occurred when the temperature was reduced is described above and explained as an increase in the ratio k_3/k_2.

Ethyl bromide is liberated when TEAB is added to an equimolar solution of acrylonitrile and triethylphosphite in DMF.

$$(C_2H_5O)_3\overset{\oplus}{P}-CH_2-\overset{\ominus}{C}HCN \;\;\overset{Br^{\ominus}}{\rightleftharpoons}\;\; (C_2H_5O)_3\overset{\oplus}{P}-CH_2\overset{|}{C}HCN \overset{Br}{\underset{}{}}$$

$$\longrightarrow (C_2H_5O)_2\overset{O}{\overset{\|}{P}}-CH_2\overset{\ominus}{C}HCN + C_2H_5Br$$

The cation may actually be destroyed by the salt. In the case of the amine only the second alternative is feasible.

Lithium chloride decreases the polymerization rate and is thought to form a contact ion pair with the carbanion.

As well as methacrylonitrile, the Russian group Kotchetov, Berlin and Yenikolopyan[18–21] also studied the polymerization of acrylonitrile by triethylphosphine at 40 °C. Addition of water reduced the rate of polymerization, the molecular weight of the polymer formed and caused the solution conductance to increase parallel with conversion of monomer to polymer. The maximum conductance, at the end of the polymerization, was directly proportional to the amount of water added. A rise in conductance is to be expected if zwitterions form and react with water. I.e.:

$$R_3\overset{\oplus}{P}\text{+}CH_2CHCN\text{+}\overset{\ominus}{C}H_2\overset{\ominus}{C}HCN + H_2O \rightarrow R_3\overset{\oplus}{P}\text{+}CH_2CHCN\text{+}_n CH_2CH_2CN$$
$$OH^{\ominus}$$

A kinetic analysis based on four processes, initiation, propagation, termination on water and spontaneous termination, led to the following expression for the initial rate of polymerization.

$$-\frac{d[M]}{dt_0} = \frac{k_i k_p [M]_0^2 [I]_0}{k_t [H_2O]_0 + k_t^1}$$

By plotting log $d[M]/dt_0$ vs. $\log[I]_0$ it could be shown that the order with respect to initiator is indeed one. However, the order with respect to monomer depends on the polymerization solvent being above two in solvents with dielectric constant below the monomers and only equal to two when dielectric constants of solvent and monomer are equal.

The authors suggest that one or more of the elementary rate constants are influenced by changes in dielectric constant. In THF the polymer molecular weight at a fixed conversion is independent of $[I]_0$ but proportional to $[M]_0^2$. Dimethylformamide and acrylonitrile have the same dielectric constant, and in this solvent the molecular weight is proportional to $[M]_0$. With THF it seems that k_p/k_t^1 is a function to $[M]_0$. Since THF and the monomer have very different dielectric constants, the dielectric constant of the polymerizing mixture will be proportional to the initial monomer concentration.

From molecular weight-conversion data the authors determined values of k_p/k_t^1 for various monomer concentrations and hence dielectric constants. Values of $k_i k_p/k_t^1$ obtainable from the expression for $d[M]/dt_0$ permit calculation of k_i. Both k_i and k_p/k_t^1 were shown to increase with $[M]_0$.

A major weakness of this work is the failure to positively identify the termination processes. Ogawa and Jaacks present good evidence for transfer to monomer in acrylonitrile polymerizations at 40 °C. Under these circumstances it is difficult to accept all the conclusions of the Russian group.

Jaacks, Eisenbach, and Kern[22] conclude that macrozwitterions are not formed when acrylonitrile is polymerized by triphenylphosphine in DMF at 30 °C. After fractionation of the polymer, phosphorus was found to be present solely as a 1 : 1 phosphine-monomer adduct C.

$$(C_6H_5)_3\overset{\oplus}{P}-CH_2CH_2CN \quad X^{\ominus}$$
$$C$$

Spectrophotometric determination of the chain end groups produced by termination with dinitrobenzoyl chloride showed that the growing carbanion and phosphonium ion concentrations were equal.

$$\ldots-CH_2-\underset{H}{\overset{CN}{\underset{|}{\overset{|}{C}}}}{}^{\ominus} + \underset{NO_2}{\underset{|}{\overset{O}{\underset{\|}{C}}-Cl}}\text{-}\underset{NO_2}{\text{Ar}} \longrightarrow \ldots-CH_2-\underset{H}{\overset{CN}{\underset{|}{C}}}-\underset{O}{\overset{\|}{C}}-\underset{NO_2}{\text{Ar}}\text{-}NO_2 + Cl^{\ominus}$$

The following polymerization mechanism is suggested:

$$(C_6H_5)_3P + H_2C=CHCN \longrightarrow (C_6H_5)_3\overset{\oplus}{P}-CH_2\overset{\ominus}{C}HCN$$

$$(C_6H_5)_3\overset{\oplus}{P}-CH_2\overset{\ominus}{C}HCN + H_2C=\underset{H}{\overset{CN}{\underset{|}{C}}} \longrightarrow (C_6H_5)_3\overset{\oplus}{P}-CH_2CH_2CN + H_2C=\overset{\ominus}{C}-CN$$

$$H_2C=\overset{CN}{\underset{|}{C}}\left[CH_2-\overset{CN}{\underset{|}{CH}}\right]_n CH_2\overset{\ominus}{C}HCN + H_2C=CHCN \longrightarrow H_2C=\overset{CN}{\underset{|}{C}}\left[CH_2-\overset{CN}{\underset{|}{CH}}\right]_{n+1} CH_2\overset{\ominus}{C}HCN$$

$$H_2C=\overset{CN}{\underset{|}{C}}\left[CH_2-\overset{CN}{\underset{|}{CH}}\right]_n CH_2\overset{\ominus}{C}HCN + H_2C=CHCN \longrightarrow H_2C=\overset{CN}{\underset{|}{C}}\left[CH_2-\overset{CN}{\underset{|}{CH}}\right]_n CH_2CH_2CN + H_2C=\overset{\ominus}{C}-CN$$

The equivalence of phosphonium salt and carbanion concentrations rules out termination by proton donation from a strong acid.

The comparatively large energy requirement of charge separation is thought to make proton removal more favourable. Models showed that when the chain DP was one or two, solvent could not separate the charges and diminish the attractive force between them. The bulk of the triphenylphosphonium ion will make its solvation by DMF or the anionic end of another zwitterion difficult. Proton removal leads to conditions much more favourable to chain growth. The cation and anion can be separated by solvent molecules. Vapour pressure osmotic measurements had shown that $(Ph_3P^{\oplus}CH_2CH_2CN)Br^{\ominus}$ was dissociated to a significant degree in DMF so that chain growth is likely to be primarily at free ions.

However, the authors suggest that vinyl monomers which do not possess acidic protons may be susceptible to polymerization via zwitterion intermediates.

Acrylic Acid

Saegusa et al.[23] found that low molecular weight polymer formed when acrylic acid was treated with triphenylphosphine. The polymerization was very slow, 5 mole-% Ph_3P in bulk monomer, held at 100 °C, gave a 66% yield of polymer after 300 h. NMR and IR spectra indicated that the polymer had the structure of poly(β-propiolactone)[1], the polyester of 3-hydroxypropionic acid. Evidently the carbanion formed rearranges to a carboxylate ion before addition of the next monomeric unit.

I.e.:

$$(C_6H_5)_3\overset{\oplus}{P}CH_2-\overset{\ominus}{C}H-COOH \rightarrow (C_6H_5)_3\overset{\oplus}{P}CH_2CH_2CO_2^{\ominus} \text{ etc.}$$

The polymer molecular weight was determined as 1560 by vapour pressure osmometry. The protons of the $(C_6H_5)_3P^{\oplus}$-group were clearly evident in the NMR spectrum of the polymer. Given the molecular weight of 1560, the ratio of the integrated signals of the phenyl and chain protons indicated that there was one phosphonium group per chain.

Washing with 1 N Na_2CO_3 reduced the intensity of the phenyl proton absorption and the molecular weight of the polymer. Apparently, triphenyl phosphonium groups are lost. Saegusa put forward the following scheme for the polymerization (see page 64).

After washing with sodium carbonate, chains terminated by acryloyl-groups will not longer be paired with $Ph_3P^{\oplus}H$ ions. Addition of the integrated signals of vinyl and phenyl protons and comparison with chain protons in the NMR spectrum of the washed polymer indicated a molecular weight of 1210, in close agreement with the VPO values (1240).

Clearly, carbanionic macrozwitterions are not likely to be stable when they are generated from a monomer with a proton transferring group. This study of the polymerization of acrylic acid has been included for the sake of completeness and to illustrate an intelligent approach to the problem of identifying the polymer molecules formed by covalent initiators.

---

[1] Systematic name: poly(oxy-1-oxotrimethylene)

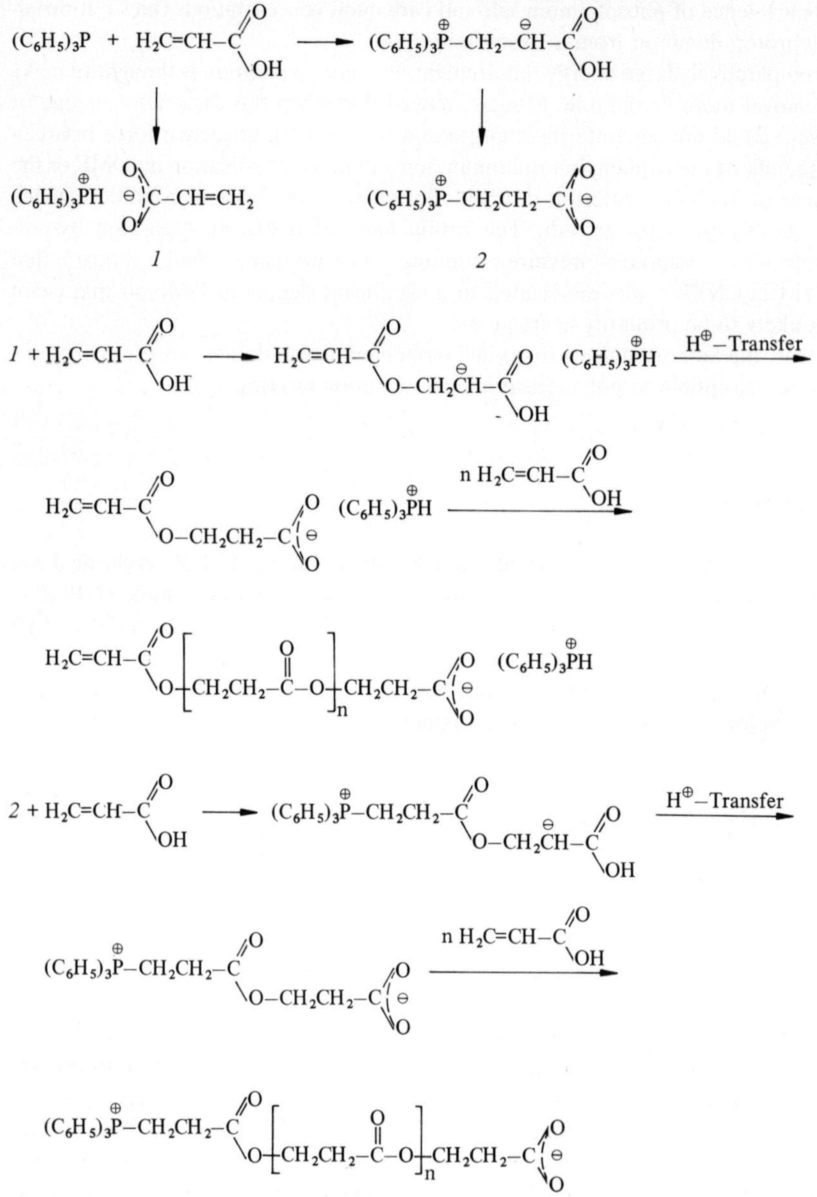

Nitroethylene

Under the extreme conditions employed by Saegusa acrylic acid will polymerize in the presence of pyridine. However, nitroethylene is the least reactive of the olefins so far polymerized by this amine in solution at room temperature.

Initial work was carried out by Vofsi and Katchalsky[24] at 20 °C in ethyl methyl ketone. They found that the rate of polymerization was second order with respect to monomer, consistent with a three step kinetic scheme:

Bimolecular initiation: $M + I \xrightarrow{k_i} {}^{\oplus}IM^{\ominus}$

Propagation: ${}^{\oplus}IM_nM^{\ominus} + M \xrightarrow{k_p} {}^{\oplus}IM_{n+1}M^{\ominus}$

Unimolecular termination: ${}^{\oplus}IM_n^{\ominus} \xrightarrow{k_t} I + P$

$$-\frac{d[M]}{dt} = \frac{k_i k_p}{k_t}[I][M]_0$$

A strong dependance of the degree of polymerization on $[M]_0$ is implied, but was not found by the authors. This discrepancy is reconciled by postulating transfer of initiator from chain end to monomer

$${}^{\oplus}IM_n^{\ominus} + M \xrightarrow{k_{tr}} P + {}^{\oplus}IM^{\ominus}$$

Hence D.P. $= \dfrac{k_p[M]}{k_t + k_{tr}[M]}$

The slight dependance of D.P. on $[M]_0$ is thus understandable if $k_{tr}[M]$ is large compared with k_t.

Values for k_p/k_t and k_p/k_{tr} are obtained by plotting 1/DP versus 1/[M]. From the ratios k_p/k_t and $k_i k_p/k_t$, k_i is available.

$\dfrac{k_p}{k_t} = 1.22 \times 10^2 \; l \cdot mol^{-1}$; $\quad \dfrac{k_p}{k_{tr}} = 17.4$

$\dfrac{k_{tr}}{k_t} = 7.0 \; l \cdot mol^{-1}$; $\quad \dfrac{k_i k_p}{k_t} = 4.7 \; mol^2 \cdot l^{-2} \cdot s^{-1}$

and hence $k_i = 3.7 \times 10^{-2} \; mol \cdot l^{-1} \cdot s^{-1}$

The value for $k_i k_p/k_t$ is an average over several batches of monomer, considerable variability was found between batches. UV spectroscopy showed that the polymer was saturated, suggesting that if termination is monomolecular macro rings are formed.

At 20 °C k_p/k_{tr} is relatively small and polymer molecular weights are limited to 2000. Vofsi, Katchalsky, and Grodzinsky[25] subsequently investigated the polymerization at temperatures as low as −104 °C in THF and −75 °C in DMF, temperatures unattainable with ethyl methyl ketone. Polymers with molecular weights as high as 300 000 were obtained.

In DMF at 20 °C second order kinetics are again observed. However at −50 °C the polymerization is first order. At −30 °C intermediate orders are found, at low monomer concentrations approaching second order, high concentrations, first order. This behaviour could be described by the following empirical equation:

$$\left(\frac{\Delta[M]}{\Delta t}\right)_{max} = \frac{a[I]_0[M^2]}{b + [M]}$$

On integration this yields:

$$b\left(\frac{1}{[M]} - \frac{1}{[M]_0}\right) + \ln([M]_0/[M]) = a[I]_0(t - t_0)$$

From their experimental results the authors determined values for a and b at 22.5, −0.5, −27.5, and −50.5 °C.

In THF the polymerization is second order at +25 °C and first order at −104 °C. At −75 °C in DMF and −104 °C in THF there is a linear relationship between degree of conversion and molecular weight. Thus, it seems that larger activation energies are associated with transfer and termination than with initiation or propagation. At low enough temperatures the polymerization can be described as an ideal "living" type.

If allowance is made for a significant utilisation of initiator the original kinetic equation describing polymerization expands to

$$-\frac{d[M]}{dt} = \frac{k_i(k_p + k_{tr})[I]_0[M]^2}{k_t + k_i[M]}$$

This is seen to be identical to the empirical equation put forward to describe the relation between $(\Delta[M]/\Delta t)_{max}$ and $[M]$, hence a is the sum of k_p and k_{tr} and b the ratio k_t/k_i.

Temperature in °C	$a[I]_0$ [a]	b [a]
22.5	5.2	14.5
− 0.5	2.3	3.2
−27.5	0.67	0.2
−50.5	0.23	0

[a] for $[I]_0 = 3.7 \times 10^{-4}$

Clearly k_t is more influenced by changes in temperature than k_p. The changes in polymerization order can, therefore, be understood.

If $k_t > k_i[M]$

$$-\frac{d[M]}{dt} = \frac{k_i(k_p + k_{tr})}{k_t}[I]_0[M]^2$$

This condition is satisfied at low initial monomer concentrations, high conversions and high temperatures.

If $k_t < k_i[M]$

$$-\frac{d[M]}{dt} = (k_p + k_{tr})[I]_0[M]$$

This condition is satisfied at high initial monomer concentrations, low conversions and low temperatures.

Using the expression $\overline{DP} = k_p/(k_{tr}[M] + k_t)$ and molecular weight-temperature data, the authors derive activation energies (A.E.) for propagation and transfer:
A.E.$_{prop}$ = 5 kcal · mol^{-1}; A.E.$_{transfer}$ = 12 kcal · mol^{-1}

With the data available, only the difference between the termination and initiation activation energies could be calculated

$A.E._{term} - A.E._{init.} = 13$ kcal · mol^{-1}

The overall activation energy is formally negative.

The general conclusion that termination/transfer reactions with higher activation energies than those of propagation become negligible at low termperatures, is logical. However, the authors attempts to define the chain breaking reactions are less convincing. In a later review Grodzinsky[26] puts forward a rather improbable three centre mechanism for chain transfer:

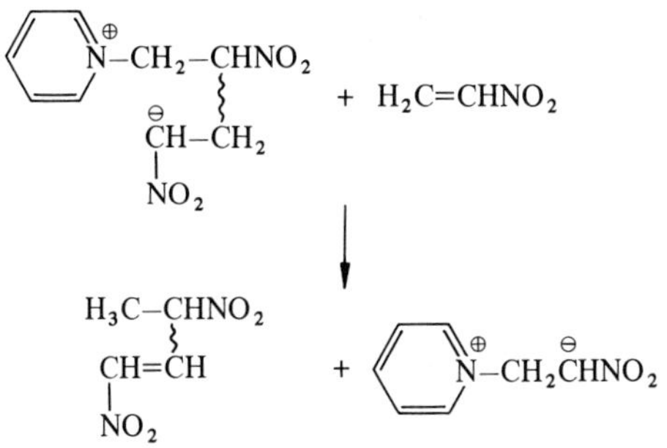

[Scheme showing three-centre chain transfer mechanism with pyridinium–CH$_2$–CHNO$_2$ / CH–CH$_2$ / NO$_2$ + H$_2$C=CHNO$_2$ → H$_3$C–CHNO$_2$ / CH=CH / NO$_2$ + pyridinium–CH$_2$CHNO$_2$]

He claims to have detected terminal unsaturation in polymers of 2-nitropropene[27], although Vofsi and Katchalsky[24] state in their paper that nitroethylene polymers are almost certainly wholly saturated molecules.

The mechanism proposed by Jaacks, Eisenbach, and Kern for chain transfer in the polymerization of acrylonitrile may well apply here.

2,4,6-Trinitrostyrene

2,4,6-Trinitrostyrene is another nitroolefin which appears to be capable of forming zwitterions with tertiary amines. Butler and Sivaramakrishnan[28] were unable to induce polymerization with a variety of ionic and free radical initiators. However, 24 h after triethylamine was added to a concentrated monomer solution, a limited yield of red-brown polymer was obtained. Polar solvents gave the highest yields. The following mechanism is suggested.

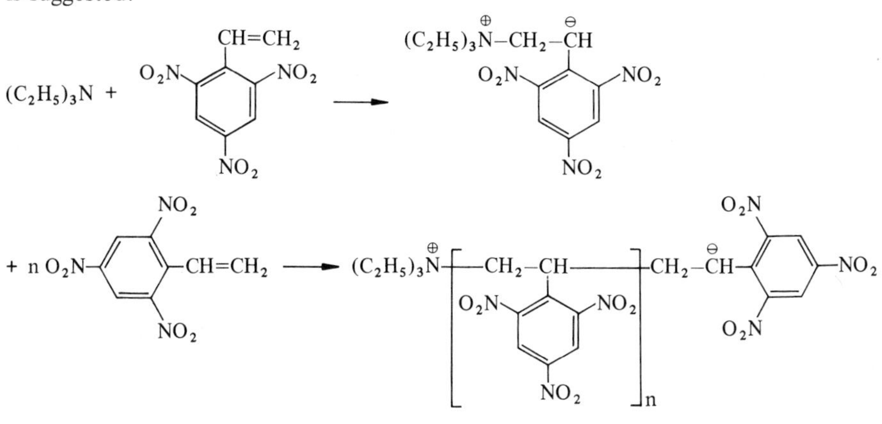

[Scheme showing $(C_2H_5)_3N$ + 2,4,6-trinitrostyrene → zwitterion $(C_2H_5)_3\overset{\oplus}{N}$–CH$_2$–$\overset{\ominus}{C}$H–Ar, then + n monomer → $(C_2H_5)_3\overset{\oplus}{N}$–[CH$_2$–CH(Ar)]–CH$_2$–$\overset{\ominus}{C}$H–Ar polymer]

It had been shown that a secondary amine would add across the double bond.

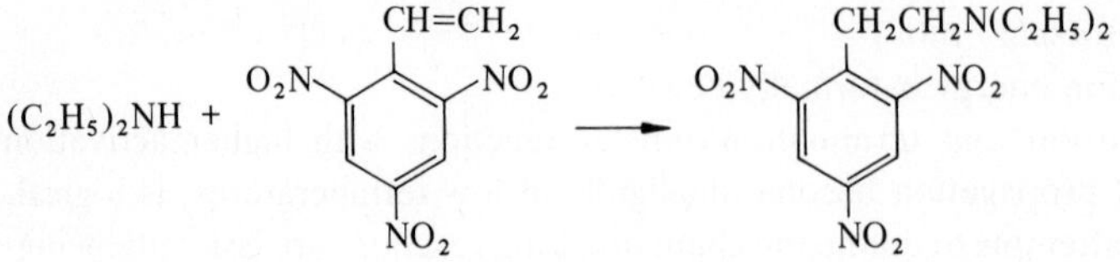

$(C_2H_5)_2NH\ +$ [2,4,6-trinitrostyrene] $\longrightarrow$ [2,4,6-trinitro-β-(diethylamino)ethylbenzene]

2-Vinyl- and 4-vinylpyridines and 4-dimethylaminostyrene copolymerize with 2,4,6-trinitrostyrene without added initiator. The authors conclude that zwitterion formation is the first step. These copolymerizations seem to be analogous to those recently investigated by Saegusa and to be described later.

Diethyl methylenemalonate

Diethyl methylenemalonate is one of the most electrophilic vinyl monomers for which evidence exists of macrozwitterion formation with Lewis bases. Hopff, Lüssi, and Allisson[29] carried out the first systematic work. They showed that diethyl methylenemalonate could be polymerized by amines, different types of amine having different initiating powers.

Rapid polymerization followed when the monomer was added to tertiary amines and phosphines such as triethylamine, triethylphosphine, N-ethylpiperidine, or pyridine. However, polymerizations initiated by tertiary amines with bulky substituents took hours rather than minutes. Primary and secondary amines were even less effective, whereas aromatic amines such as aniline and N, N-dimethyl-p-toluidine did not cause polymerization.

The authors found that unless a strong acid (sulfuric or benzene sulfonic acid) was added to the monomer it could not be stored without "spontaneously" polymerizing.

Jaacks and Franzmann[30] studied the polymerization of this monomer by triphenylphosphine solely with the aim of demonstrating the existence of macrozwitterions. To facilitate end group analysis, a high ratio of initiator to monomer was used to give relatively low polymer molecular weight. The UV spectrum of the polymer obtained was identical to that of the model compound D.

$$(C_6H_5)_3\overset{\oplus}{P}-CH_2CH(COOC_2H_5)_2\ Br^{\ominus}$$
$$D$$

UV spectroscopy shows that quaternary phosphonium ions are present but, of course, does not prove that they are joined to polymer chains. The authors fractionated their polymers with a THF/water solvent-non solvent combination. They had found that the model compound described above could be quantitatively separated from phosphorus free poly(methylenemalonic ester). After fractionation the phosphorus content fell, but when the molecular weight was determined by vapour pressure osmometry, it was found that there was approximately one phosphonium group per chain. The absorption coefficient of the model compound was used to calculate the phosphorus content of the product.

Low molecular weight polymer (<2000) was removed by GPC and the phosphorus content and molecular weight of the remainder determined. Again it could be shown that there was one phosphonium group per chain. The authors conclude that macrozwitterions are the chain carriers in this polymerization.

Alkyl 2-Cyanoacrylates

Recently Johnston and Pepper[31-35] investigated the polymerization of various esters of 2-cyanoacrylic acid in tetrahydrofuran. As with diethyl methylenemalonate traces of acid had to be added to the monomers if uncontrollable "spontaneous" polymerization was to be avoided.

Phosphine polymerizations were much simpler than those initiated by amines. UV spectra of polymer solutions showed clearly that macrozwitterions had been formed.

$$R_3P + n\ H_2C=C(CN)(COOR) \longrightarrow R_3P^{\oplus}\text{–}[CH_2\text{–}C(CN)(COOR)]_{n-1}\text{–}CH_2\text{–}C^{\ominus}(CN)(COOR)$$

If acid was added there was a marked decrease in absorbance around 250 nm. The difference spectrum, a plot of fall in absorbance versus wavelength, was found to be identical to the spectrum of sodium butyl cyanoacetate (Na$^{\oplus}$ $\overset{\ominus}{C}$H(CN)COOBut). If the initiator was triphenylphosphine the absorption remaining after addition of acid was identical to triphenyl-methylphosphonium bromide. Repetitive isolation and purification of the polymer did not reduce the intensity of this absorption. Assuming that the absorption coefficients of the model compounds and chain ions were identical it could be shown that molarities of polymer, cation, and anion were equal to the initiating phosphine concentration.

The conductivity of polymer solutions increased when benzoyl chloride was added until polymer and benzoyl chloride were equimolar. A rise in conductivity is to be expected if zwitterions are present and interact with benzoyl chloride to form a polymeric salt.

$$(C_2H_5)_3P^{\oplus}\text{–}[CH_2\text{–}C(CN)(COOR)]_n\text{–}CH_2\text{–}C^{\ominus}(CN)(COOR) + C_6H_5COCl$$

$$\longrightarrow (C_2H_5)_3P^{\oplus}\text{–}[CH_2\text{–}C(CN)(COOR)]_n\text{–}CH_2\text{–}C(CN)(COOR)\text{–}COC_6H_5\ \ Cl^{\ominus}$$

Overall polymerization rates which are equated with the rate of propagation were first order in monomer and very fast. At −78 °C the rate constants fell between 1.0 and $3.4 \times 10^5 \, l \cdot mol^{-1} \cdot s^{-1}$. Activation energies were very small, 2.2 and 5.5 kJ · mol^{-1} for ethyl (ECA) and butyl (BCA) cyanoacrylates, respectively. Of a range of ammonium, phosphonium, and alkali metal salts only lithium bromide significantly reduced the rate of polymerization. Ogawa and Romero[16] found the rate of acrylonitrile polymerization increased by the presence of an ammonium salt but reduced by lithium chloride. There may be a specific interaction between cyano substituted carbanions and the lithium cation.

On the basis of the very large k_p values and great dilution of monomer solutions the authors suggest that monomer addition takes place at free ions.

Monomer added at the end of polymerizations linked to chains already present and molecular weights were determined solely by the monomer/initiator ratio. However, dispersity ratios, M_w/M_n, do not approach the values (1.02–1.05) attained in truly ideal living polymerizations. The molecular weight distributions of polymer formed when monomer is added slowly to initiator solutions were even broader suggesting that initiation is slower than propagation.

Johnston and Pepper conclude that phosphines initiate near ideal living polymerizations. However, when the authors turned to amine initiators they found that, although macrozwitterions were formed, the polymerization kinetics were very different. At comparable reagent concentrations room temperature rates were at least one thousand times slower, but paradoxically increased as temperature was reduced. Arrhenius plots indicated that by −100 °C amine and phosphine polymerization rates would be equal. Polymer molecular weights were much higher than would have been expected had initiation been complete, and were uninfluenced by polymerization conditions. It is believed that molecular weights are determined by traces of weak acid transfer agents present in the monomer.

Polymerizations initiated by pyridine, while not exactly first order in monomer over the whole time course, approach first order behaviour very closely at high conversion. Despite this, the external order with respect to monomer was found to be two.

Two experimental results lead the authors to conclude that reaction between pyridine and cyanoacrylate is reversible.

If monomer and a comparatively high concentration of pyridine are mixed at room temperature, monomer equivalent to the pyridine concentration does not immediately react. (Acid, a fraction of the pyridine concentration, had to be added to suppress propagation.) At −95 °C there is little difference between the rates of phosphine and pyridine initiated polymerizations. Phosphine reacts rapidly with cyanoacrylate, so must pyridine. That an equivalent of monomer does not react very rapidly at room temperature suggests that this exothermic reaction is reversible.

Conductivity measurements support this view and give some indication of the influence of polymerization variables on the equilibrium constant.

I.e.:

$$\text{Pyridine} + H_2C=C\begin{array}{c}CN\\|\\COOR\end{array} \rightleftharpoons \text{Pyridine}^{\oplus}-CH_2-C^{\ominus}\begin{array}{c}CN\\|\\COOR\end{array}$$

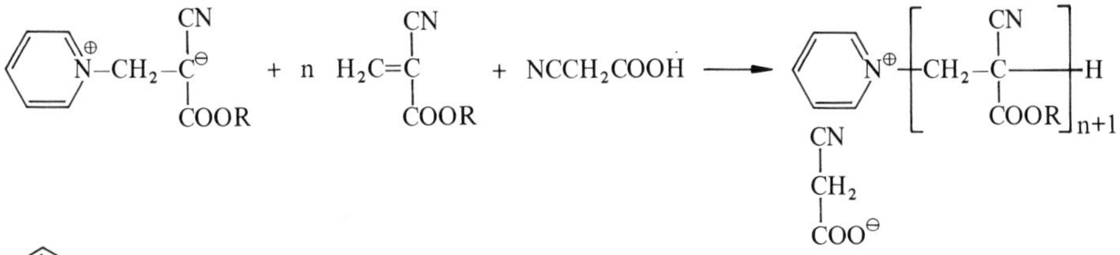

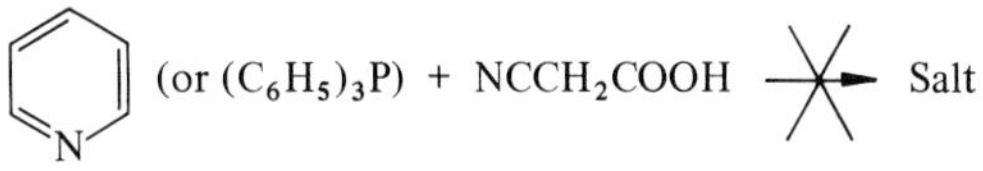

Macrozwitterion Polymerization

[Structure: Pyridine-N⁺-CH₂-C⁻(CN)(COOR) + n H₂C=C(CN)(COOR) + NCCH₂COOH → Pyridine-N⁺-[CH₂-C(CN)(COOR)]ₙ₊₁-H with CH₂-COO⁻ group attached to N via CH₂-C(CN)]

Pyridine (or $(C_6H_5)_3P$) + $NCCH_2COOH$ ⇏ Salt

$R = CH_3;\ C_2H_5;\ (CH_2)_3CH_3$

$[Py] = [(C_6H_5)_3P] = 13\ \text{mmol}\cdot l^{-1}$

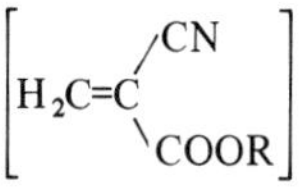

$[H_2C=C(CN)(COOR)] = 20\ \text{mmol}\cdot l^{-1};\ [NCCH_2COOH] = 65\ \text{mmol}\cdot l^{-1}$

In the case of pyridine the rate of conductivity rise increases as the temperature is reduced and is influenced by the length of the monomer alkyl chain. Time to half conductance rise is used as a measure of rate and at room temperature values for $t_{1/2}$ are: Butyl ester: 13.8 s, ethyl ester: 8.3 s, methyl ester: 1.4 s

Ethyl 2-cyanoacrylate polymerises more rapidly than its butyl counterpart, and the conductivity measurements suggest that this is because the equilibrium concentration of the ethylpyridinium betaine is greater.

The conductivity rise with the phosphine is too fast to be measureable at any temperature.

The authors put forward the following polymerization scheme:

$$Py + CA \rightleftharpoons {}^{\oplus}PyCA^{\ominus}$$
$${}^{\oplus}PyCA^{\ominus} + CA \xrightarrow{k_i} {}^{\oplus}PyCA\ CA^{\ominus}$$
$${}^{\oplus}PyCA\ CA^{\ominus} + CA \xrightarrow{k_p} {}^{\oplus}PyCA\ CA\ CA^{\ominus}$$
$${}^{\oplus}Py(CA)_n CA^{\ominus} + CA \xrightarrow{k_p} {}^{\oplus}Py(CA)_{n+1}CA^{\ominus}$$

Py = pyridine, CA = cyanoacrylate

Study of the polymerization kinetics in the presence of strong acid and the dependence of the initial rate of conductivity rise on the second power of monomer concentration suggest that monomer is involved twice in the initiation process.

Benzyldimethylamine and triethylamine initiated polymerizations are even slower ($k_{obs}/[I]_0$ for pyridine is larger) and have a more pronounced "negative" activation energy. The order with respect to the butyl monomer is internally first at all conversions but externally third. The characteristics of the polymerization of the ethyl monomer are even more unusual. The Arrhenius plot has a break at 0 °C, there being a very marked fall off in rate above this temperature. Conversion-time curves accelerate to a first order monomer dependence and the external order is second.

Added salt can increase the rates of aliphatic amine polymerization more than ten fold. Salt effects are generally more marked under conditions which promote slow polymerization. Even lithium bromide increases the polymerization rate and at low temperatures in the presence of this salt, initiation is complete and polymer molecular weights are determined by monomer/initiator ratios. The time-conversion curves for the butyl monomer at low temperatures in the presence of lithium bromide are not first order throughout, but like the ethyl ester show an initial acceleration.

The authors believe that amines initiate more slowly than phosphines. Phosphines are more readily quaternized by electrophilic carbon, and charge separation does not present such a barrier to chain initiation.

The amine initiation rate depends strongly on monomer concentration. This dependence coupled with a *very* fast rate of propagation at room temperature ensure that kinetic chain lengths are long. The rate of chain formation will fall off rapidly so that the growing chain population is essentially constant over the majority of conversion. Any change which increases the ratio of the rate of initiation to the rate of propagation will cause a greater change in chain concentration during polymerization and the conversion curves will become more obviously "s" shaped.

Molecular weights of polymers, as with the initiator pyridine, are very high, unaffected by polymerization conditions and presumably determined by the level of weak acids present. Only at low temperatures, in the presence of lithium bromide, are sufficient chains initiated for molecular weights to be uninfluenced by adventitious chain transfer agents.

The conductivity technique used to study the pyridine initiation process could not be applied to aliphatic amines because they react appreciably with weak acids. However, some information was gained from experiments in which further monomer was added to completed polymerizations.

Clearly after the initial polymerization of experiment No. 2 no free amine remained. Yet polymer molecular weight is such that only a minute fraction of the amine could be linked to long polymer chains. The authors suggest that the amine is quantitatively

Table 2. Fate of monomer added to polymer solutions

Exp. No.	First polymerization[a]	Monomer added	Result	Salt added	Result
1	BCA/BzMe$_2$N	ECA	ECA polymerized	–	–
2	ECA/BzMe$_2$N	BCA	No polymerization	Yes	BCA slowly polymerized
3	ECA/Pyridine	BCA	BCA polymerized	–	–

[a] BzMe$_2$N = benzyldimethylamine

converted to a stable betaine unable to react with butyl ester monomer and only reversibly, judging by the negative activation energy of polymerization, with further ethyl 2-cyanoacrylate. By contrast, betaines formed from BDMA/BCA or pyridine/ECA are inferred to be less stable, probably in equilibrium with base and monomer and able to directly initiate polymerization of either ethyl or butyl 2-cyanoacrylate.

It would seem that added salt can increase the reactivity of the ECA betaine.

The following *general* mechanism is put forward to explain these observations.

$$R_3N + CA \rightleftharpoons R_3N^\oplus CA^\ominus$$

$$R_3N^\oplus CA^\ominus + CA \rightleftharpoons R_3N^\oplus CA\ CA^\ominus$$

$$R_3N^\oplus CA\ CA^\ominus + CA \xrightarrow{k_i} R_3N^\oplus CA\ CA\ CA^\ominus$$

$$R_3N^\oplus (CA)_n CA^\ominus + CA \xrightarrow{k_p} R_3N^\oplus (CA)_{n+1} CA^\ominus$$

So far as could be determined, the only difference between polymerizations initiated by tribenzylamine and other aliphatic amines was in rate, tribenzylamine initiated polymerizations being two orders of magnitude slower.

The authors conducted a limited number of experiments in solvents other than THF. In methylene chloride rates were too slow to be measured accurately, in the nitro solvents, nitromethane and nitrobenzene, too fast. There was little difference in rate between polymerizations conducted in pure THF or THF/methylene chloride mixtures.

Methylenemalononitrile

Methylenemalononitrile would seem to be the vinyl monomer most likely to form macrozwitterions in the presence of covalent bases. However, its homopolymerization has not been studied, presumably because of its extreme reactivity and the insolubility of the polymer.

Stille, Oguni, and Kamachi[36] found that when methylenemalononitrile and tetrahydrofuran were mixed each polymerized independently of the other. The rate of disappearance of methylenemalononitrile increased when the initial proportion of tetrahydrofuran was increased and the molecular weight of both polymers increased as the concentration of the other was reduced. This strongly suggests mutual initiation. Methylenemalononitrile polymerized rapidly and precipitated. It is suggested that carbanions, buried in precipitated polymer could be fed by monomer from solution and not hinder subsequent cationic polymerization of the ether. Stille et al. put forward the following mechanism (see page 74).

Models showed that cyclization of the alkene ether adduct was sterically impossible. Cyclization after addition of the second monomeric unit is possible but apparently cannot compete with propagation. Cyclization to ring sizes greater than nine becomes progressively less favourable.

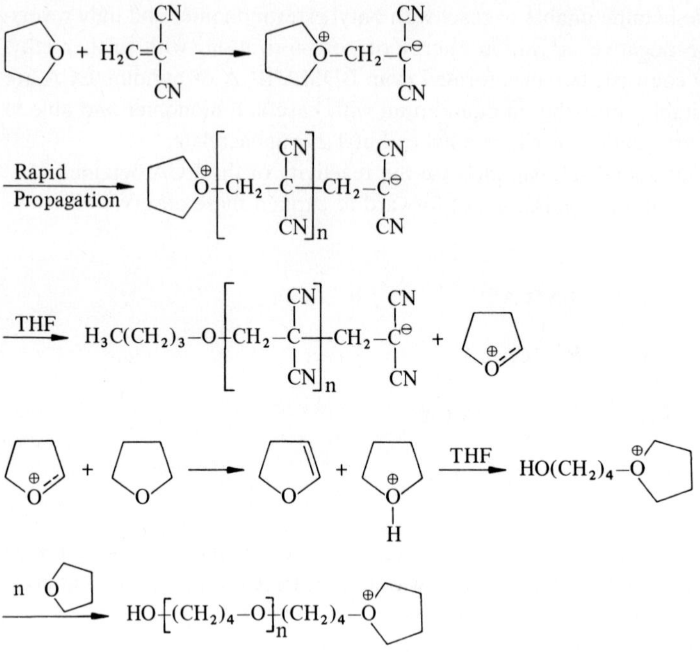

Methyl Vinyl Ether

There is only one report in the literature of a vinyl polymerization where the chain carriers could be zwitterions with cationic active centres. Stille[37] found that high molecular weight polymer ($M_n = 10^5$) formed when a catalytic amount of 2,3-dichloro-5,6-dicyano-p-benzoquinone (DDQ) or tetracyanoquinodimethane[2] was added to a vinyl ether.

Immediately after mixing a colour characteristic of a charge transfer complex was observed. It eventually disappeared at a rate dependent on the solvent, temperature, and vinyl ether. Polar solvents and higher temperatures led to faster rates as did an increase in the size of the alkyl substituent, $C(CH_3)_3 > CH(CH_3)_2 > C_2H_5$. Tetracyanoethylene also forms charge transfer complexes with these monomers, but the final products are cyclobutane 2 + 2 cyclo addition compounds rather than polymer.

When DDQ and methyl vinyl ether (1:1) were flow mixed into methanol, the 1:1 reaction product was obtained.

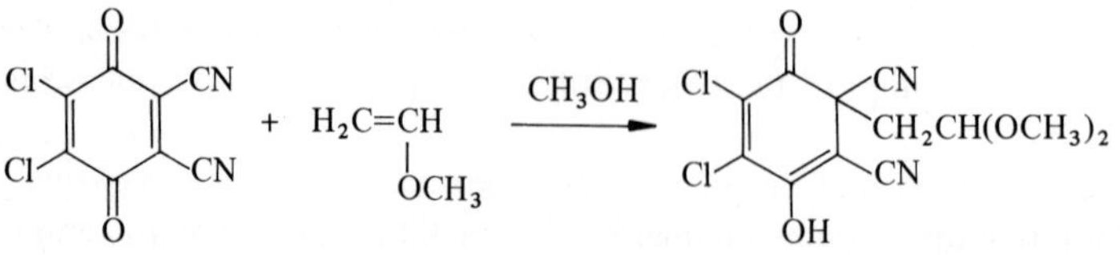

---

[2] Systematic name: 1,2,4,5-tetracyano-3,6-dimethylene-1,4-cyclohexadiene

Hence, the likely polymerization mechanism is the following one.

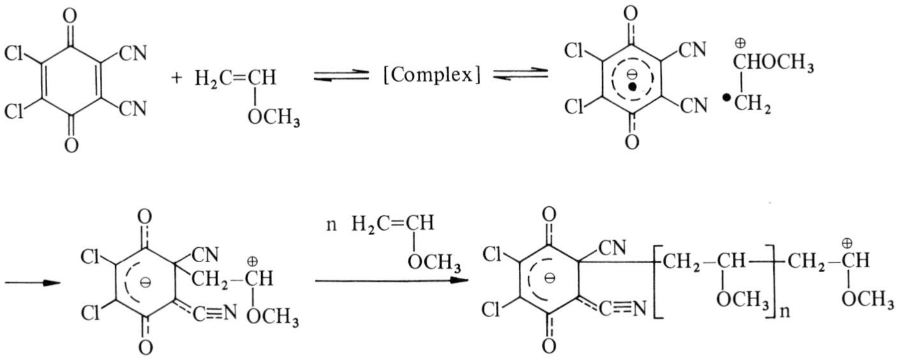

$$\text{chloranil-dicyano} + H_2C=CH(OCH_3) \rightleftharpoons [\text{Complex}] \rightleftharpoons [\text{radical anion} \cdots \overset{\oplus}{C}H(OCH_3)\text{-}\overset{\bullet}{C}H_2]$$

$$\rightarrow [\text{anion} \cdots \overset{\oplus}{C}H_2\text{-}CH(OCH_3), C\equiv N] \xrightarrow{n\ H_2C=CH(OCH_3)} [\text{anion}\equiv C \equiv N]\text{-}[CH_2\text{-}CH(OCH_3)]_n\text{-}CH_2\text{-}\overset{\oplus}{C}H(OCH_3)$$

This is a reasonable theory but an unequivocal demonstration of the presence of zwitterions is needed.

2.1.2 Carbonyl Monomers

Formaldehyde

Formaldehyde possessing a strongly polarised carbonyl bond is very susceptible to either cationic or anionic addition polymerization. With a basic initiator the active centre is an alkoxide ion in contrast to the carbanion of a vinyl monomer. The readiness with which polymerization occurs makes initiation by relatively weak covalent bases possible.

Kinetic studies, however, are complicated by the water content of the monomer. It seems to be impossible to dry formaldehyde completely.

Machacek, Mejzlik, and Pac[38], who carried out the first kinetic experiments with dibutylamine, assume that only the hydroxide ion formed by hydrolysis of the amine adds to the double bond. In diethyl ether at $-60\,°C$ conversion/time curves displayed a high internal order with respect to monomer and external first order with respect to initiator.

Enikolopyan[39] studied the polymerization of this monomer by triethylamine and the ionic salts, calcium stearate and tetrabutylammonium laurate. From kinetic arguments Enikolpyan concludes that a cocatalyst, presumably water, is necessary for amine initiated polymerization. Overall first order kinetics were found with the two salts, whereas with triethylamine the polymerization was of higher order.

Mathes and Jaacks[40] examined polymer end groups for evidence of macrozwitterion formation. Low molecular weight polymers were obtained with the aid of high initiator/monomer ratios, a high dielectric constant solvent and addition of carbon dioxide:

$$\ldots -CH_2-O-CH_2-\overset{\ominus}{O} + CO_2 \rightleftharpoons \ldots -CH_2OCH_2OC\overset{O}{\underset{O}{\diagdown}}{}^{\ominus}$$

Elemental analyses for nitrogen by the methods of Kjeldahl and Lassaigne were negative. None of the initiator, triethylamine, could be detected after the poly(oxymethylene) chain had been depolymerized.

$$(C_2H_5)_3\overset{\oplus}{N}-(CH_2O)_n CH_2O^{\ominus} \xrightarrow[\text{2. Na}_2\text{CO}_3]{\text{1. HCl}} (C_2H_5)N + n\,HOCH_2OH$$

The absence of carbonate end groups could be demonstrated by IR spectroscopy, carbonate ions would be stable in the presence of quaternary ammonium ions but not tertiary.

$$\ldots-CH_2OCH_2OCOO^{\ominus} + H\overset{\oplus}{N}R_3 \rightleftharpoons \ldots-CH_2OCH_2OCOOH$$

$$\ldots-CH_2OCH_2OH + CO_2$$

Termination of the alkoxide ions with ethyl iodide and determination of the resulting ethoxide end groups showed that the concentration of poly(oxymethylene) anions and quaternary ammonium cations would have been high enough for detection by the techniques used.

Under the experimental conditions quaternary ammonium ions would not have been subject to Hofmann degradation. The growing anions must then have been partnered almost exclusively by tertiary ammonium counter ions.

However, if the rate of propagation is much faster with quaternary ammonium cations, a small number of these ions could have been associated with the formation of a significant proportion of the polymer. A fast termination reaction is to be expected with tertiary ammonium cations.

$$\ldots-CH_2OCH_2O^{\ominus} + H\overset{\oplus}{N}(C_2H_5)_3 \rightarrow \ldots-CH_2OCH_2OH + N(C_2H_5)_3$$

The authors do not totally rule out the zwitterion machanism but clearly it cannot be the only polymerization mechanism here.

As well as amines Mathes and Jaacks investigated the polymerization with two tertiary phosphines, Ph_3P and $((CH_3)_2N)_3P$. Flame photometry showed that in this case the polymers contained significant amounts of phosphorus, one triphenylphosphine residue per fifteen chains and one tris(dimethylamino)phosphine residue per three chains.

It seems that phosphines add directly to the carbonyl bond, but the presence of chains not terminated by phosphorus indicates that there is a second initiation mechanism and/or termination-transfer processes.

The authors of the three publications, so far described, could not find any evidence for N-zwitterion formation. Künzel, Giefer, and Kern[41] believe they have. This group measured the effectiveness of a wide range of covalent bases as initiators of formaldehyde polymerization.

Most of their paper, like the majority of the literature, is concerned with amines. However, some information about the activity of Lewis bases of other elements of group five of the periodic table is presented.

Triphenylphosphine is a very effective initiator, surprisingly so, if pK_a can be taken as a guide to reactivity of bases toward formaldehyde. Compared to tributylamine it causes faster polymerization in ether and toluene (but not in acetone), despite the fact that its pK_a is many orders of magnitude smaller.

For amines of similar pK_a activity increased along the series primary, secondary, tertiary. Changes in solvent, monomer concentration and amount of adventitious carbon

Table 3. Initiator activity of Lewis basis as function of the dipole moment[a]

Initiator	Dipole Moment (D)	Conversion after 15 in %
Triphenylamine	0.26	0
Triphenylstilbene	0.57	11.2
Triphenylarsine	1.07	8.5
Triphenylphosphine	1.45	72.1

[a] $[M]_0 = 5.64$ mol·l^{-1}; $[I]_0 = 10^{-3}$ mol·l^{-1}; $T = -35\,°C$; solvent: diethyl ether

dioxide did not alter this relationship. Increasing the solvent dielectric constant, through toluene-ether-acetone, increased the polymerization rate.

Closer examination of the primary amine results, showed that n-alkyl chain length did not influence the polymerization rate. However, amines with branching on the α-carbon were completely ineffective as initiators. Branching on the β-carbon did not have such an inhibiting effect (especially so in the case of benzylamine) and there was no difference between the γ-branched amine and its straight chain analogue.

The authors suggest that an explanation for these observations is to be found in the following series of reactions:

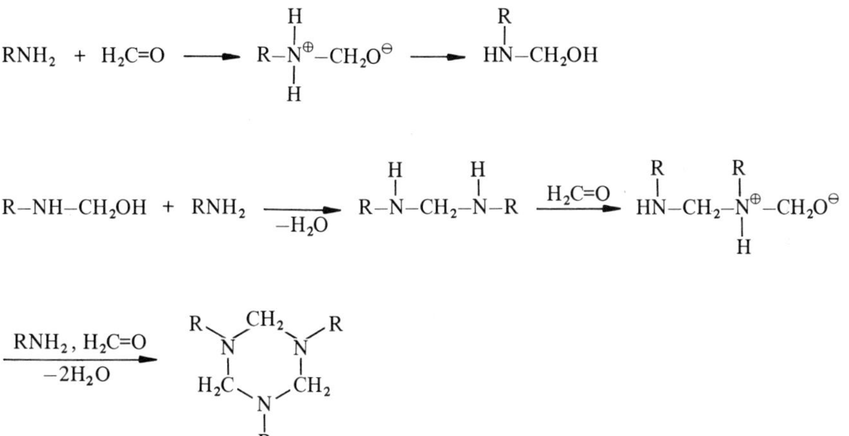

$$RNH_2 + H_2C=O \longrightarrow R-\overset{H}{\underset{H}{\overset{\oplus}{N}}}-CH_2O^{\ominus} \longrightarrow H\overset{R}{\underset{}{N}}-CH_2OH$$

$$R-NH-CH_2OH + RNH_2 \xrightarrow{-H_2O} R-\overset{H}{\underset{}{N}}-CH_2-\overset{H}{\underset{}{N}}-R \xrightarrow{H_2C=O} H\overset{R}{\underset{}{N}}-CH_2-\overset{R}{\underset{H}{\overset{\oplus}{N}}}-CH_2O^{\ominus}$$

$$\xrightarrow[-2H_2O]{RNH_2,\ H_2C=O} \quad \text{(triazine ring with R, CH}_2\text{, R, CH}_2\text{, R, CH}_2\text{)}$$

$$R_2NH + H_2C=O \longrightarrow R_2NCH_2OH$$

$$R_2NCH_2OH + R_2NH \longrightarrow R_2NCH_2NR_2 + H_2O$$

The hydroxymethyl compounds do not initiate polymerization. Only when there is a significant amount of tertiary amine can polymerization begin. Tertiary amine is formed more rapidly from secondary amines, because there is only one hydrogen atom to be replaced. Branching on the α-carbon of primary amines makes the formation of triazine sterically impossible.

It seems, therefore, that amines can add directly to the double bond of formaldehyde.

There is a rough correlation between the pK_a of an amine and its effectiveness as an initiator, but especially within the tertiary amine category there are some glaring exceptions to this rule. Interpretation of the results with tertiary amine is not complicated by formation of hydroxymethyl groups.

Pyridine (pK_a 5.2) causes more rapid polymerization than dimethyl-*tert*-butylamine (pK_a 10.5). It is difficult to reconcile this observation with the idea that the initiating species is a hydroxyl ion formed by hydrolysis of the amine. On the other hand the structure of an amine might be expected to influence its rate of addition to the carbonyl double bond. The first and rate determining step in *the formation of a long chain polymer* is apparently betaine formation:

$$R_3N + H_2C=O \rightleftharpoons R_3\overset{\oplus}{N}CH_2O^{\ominus}$$

Jaacks[40] has suggested that, while most chains initiated must be ion pairs, zwitterions could make up most of the polymer. An alkoxide ion will react rapidly with a tertiary ammonium counter ion and consequently the molecular weigth of chains initiated by hydroxide ion is probably quite low.

2.1.3 Strained Ring Monomers

Lactones

Of the cyclic monomers which polymerize following a nucleophilic attack on the ring, lactones are the most studied group which are reputed to form macrozwitterions. Polymerization is uncomplicated by side reactions, consequently macrozwitterions can be generated easily and unequivocally.

Etienne and Soulas[42] studied the polymerization of α,α-disubstituted derivatives (type E), of α,α,β-trisubstituted derivatives (type F) and of unsubstituted derivatives.

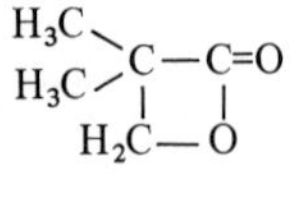

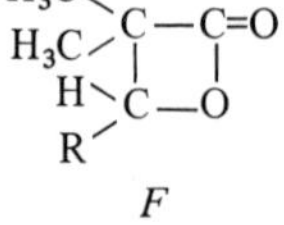

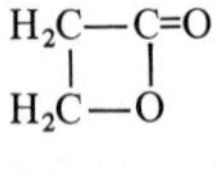

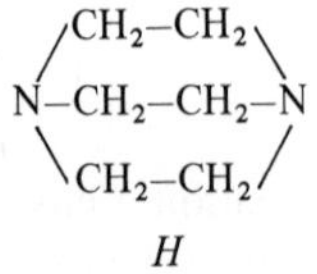

Amines like triethylenediamine (H) with readily accessible nitrogen atoms are especially effective initiators. In the presence of such amines or preformed betaines pivalolactone (E) forms high molecular weight polymer.

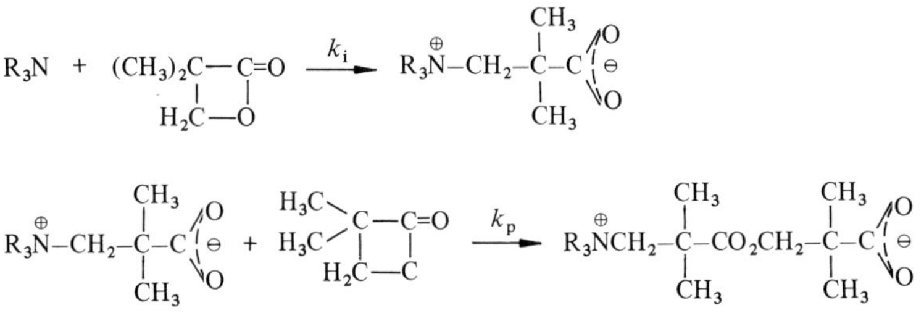

$$R_3N + (CH_3)_2C-C=O \xrightarrow{k_i} R_3\overset{\oplus}{N}-CH_2-\underset{CH_3}{\overset{CH_3}{C}}-C\overset{\nearrow O}{\underset{\searrow O}{}}{}^{\ominus}$$
$$ \underset{H_2C-O}{}$$

$$R_3\overset{\oplus}{N}-CH_2-\underset{CH_3}{\overset{CH_3}{C}}-C\overset{\nearrow O}{\underset{\searrow O}{}}{}^{\ominus} + \underset{H_2C-C}{\overset{H_3C}{\underset{H_3C}{C}}C-C=O} \xrightarrow{k_p} R_3\overset{\oplus}{N}CH_2-\underset{CH_3}{\overset{CH_3}{C}}-CO_2CH_2-\underset{CH_3}{\overset{CH_3}{C}}-C\overset{\nearrow O}{\underset{\searrow O}{}}{}^{\ominus}$$

By comparison unsubstituted lactones such as propiolactone (G) polymerize more slowly and molecular weights are lower. The authors believe propagation is slower, the carboxylate anion formed from pivalolactone ring opening being more nucleophilic because of the inductive effect of the two methyl substituents.

α,α,β-trisubstituted lactones (type F) are not polymerized by tertiary amines, but polymer does form in the presence of betaine. Steric factors are evidently responsible for this anomaly.

Mathes and Jaacks[43, 44] have carried out detailed studies of the tertiary amine and betaine initiated polymerization of propiolactone. Some of the most convincing evidence for macrozwitterions as the chain carriers in a polymerization is to be found in their work.

With both initiators, high initiator/monomer ratios, relatively low temperatures, and polar solvents were chosen to ensure that polymer molecular weights were low and the end group concentration high.

Gas chromatographic measurements indicated that when amine initiated polymers were hydrolysed with base, one amine molecule was liberated for every chain.

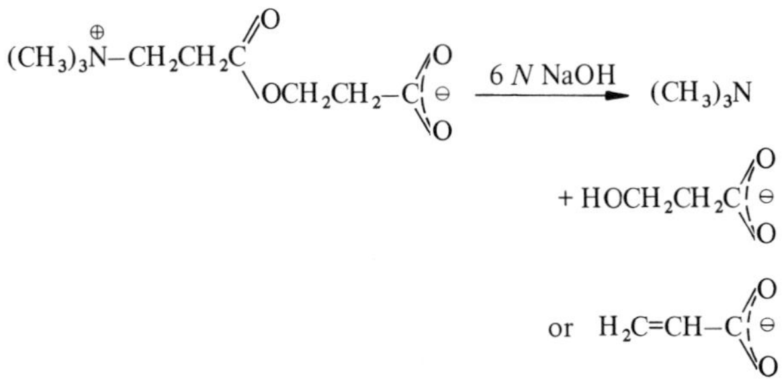

$$(CH_3)_3\overset{\oplus}{N}-CH_2CH_2C\overset{\nearrow O}{\underset{\searrow OCH_2CH_2-C\overset{\nearrow O}{\underset{\searrow O}{}}{}^{\ominus}}{}} \xrightarrow{6\,N\,NaOH} (CH_3)_3N$$

$$+ HOCH_2CH_2C\overset{\nearrow O}{\underset{\searrow O}{}}{}^{\ominus}$$

$$\text{or } H_2C=CH-C\overset{\nearrow O}{\underset{\searrow O}{}}{}^{\ominus}$$

After acid hydrolysis (quaternary ammonium ions are inert) and neutralization with sodium carbonate no amine could be detected. The polymer cannot contain free amine or tertiary ammonium ions. The only two reasonable alternatives to the zwitterionic mode of propagation can be ruled out.

I.e.:

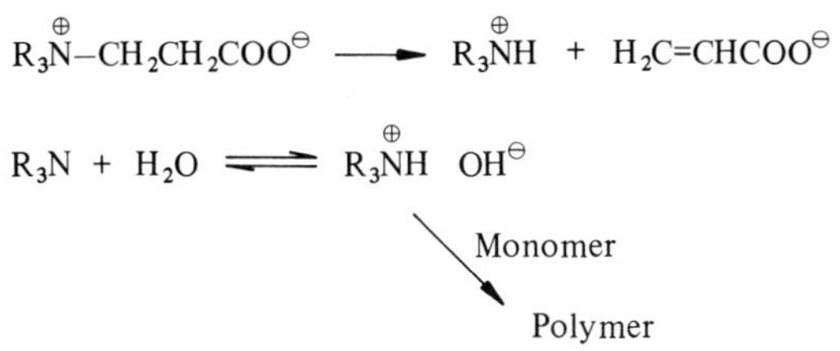

$$R_3\overset{\oplus}{N}-CH_2CH_2COO^{\ominus} \longrightarrow R_3\overset{\oplus}{N}H + H_2C=CHCOO^{\ominus}$$

$$R_3N + H_2O \rightleftharpoons R_3\overset{\oplus}{N}H\; OH^{\ominus}$$

$$\searrow \text{Monomer}$$
$$\searrow \text{Polymer}$$

A ^{1}H NMR spectrum of the model compound $(CH_3CH_2)_3\overset{\oplus}{N}CH_2CH_2COOCH_3$ $I^\ominus$ contained peaks at $\delta = 1.2$ and 3.8, which are assigned to $CH_3C-\overset{\oplus}{N}$ and $-C-CH_2-\overset{\oplus}{N}$, respectively. Similar peaks were evident in the spectrum of the polymer formed with triethylamine, their intensity corresponding to one ammonium group per chain.

In addition to the chemical tests, measurements of the dielectric constant of polymerizing solutions also suggested zwitterion formation. With amine initiators, but not sodium acetate, the dielectric constant increased steadily with conversion. An approximate chain end to end distance was calculated from the increase in dielectric constant and it corresponded closely to the value expected for a random coil.

The betaine chosen for investigation was $(CH_3)_3\overset{\oplus}{N}-CH_2COO^\ominus$. Unfortunately the solvent, ethanol, caused chain transfer.

$$\ldots-CH_2CH_2-C\underset{O}{\overset{\nearrow O}{|}}\ominus\ +\ C_2H_5OH \longrightarrow \ldots-CH_2CH_2-C\underset{OH}{\overset{\nearrow O}{}}\ +\ C_2H_5O^\ominus$$

In addition to zwitterions three other chain types were present.

The presence of carboxylate anions in the polymer was demonstrated by IR spectroscopy. The polymer and model compound $HOCH_2CH_2COO^\ominus Na^\oplus$ had identical absorptions at 6.3 μm, which disappeared on exposure to HCl-vapour. No evidence for the presence of quaternary ammonium ions can be gained from IR spectroscopy, but the presence of the only conceivable cations $H^\oplus$ and $(CH_3)_3N^\oplus-CH_2COOH$ can be ruled out. Carboxylate anions could not exist in their presence.

More information about the nature of the cation was gained from NMR spectroscopy. By reference to the model compound $(CH_3)_3N^\oplus-CH_2COO(CH_2)_9CH_3$ $Br^\ominus$ an absorption at $\delta = 3.85$ could be assigned to $^\oplus-\overset{|}{\underset{|}{N}}-CH_2-\overset{/\!\!/ O}{\underset{\backslash}{C}}$ at the cationic chain end. The relative area of this peak compared to $-CH_2-$groups in the polymer chain, suggested one ammonium ion per thirteen monomeric units. This is in fair agreement with a value of one in ten given by Kjeldahl nitrogen analysis.

The authors determined the amounts of carboxylate anion and carboxylic acid in the polymer by acid-base titration. By assuming the polymer contains an equal number of moles of ammonium cation and carboxylate anion an additional estimate of the number of monomeric units per ammonium ion was obtained. It, twelve, falls between the other two.

Dielectric and electrophoretic measurements supported the hypothesis that zwitterions were formed. Electrophoretic mobility could be demonstrated for a polymer sample after its esterification with methyl iodide.

$$(CH_3)_3\overset{\oplus}{N}-CH_2-\overset{O}{\overset{\|}{C}}-O-[CH_2CH_2-\overset{O}{\overset{\|}{C}}-O]_n-CH_2CH_2-COO^\ominus$$

$$\xrightarrow{CH_3I} (CH_3)_3\overset{\oplus}{N}-CH_2-\overset{O}{\overset{\|}{C}}-O-[CH_2CH_2-\overset{O}{\overset{\|}{C}}-O]_n-CH_2CH_2COOCH_3$$
$$I^\ominus$$

Oligomers, with n = 1 to 3, were completely separated.

A 0.25% chloroform solution of ammonium polymer had a dielectric constant 4.5% higher than the pure solvent. No change in dielectric constant could be detected when hydroxyl ion initiated polymer was dissolved in chloroform.

Having clearly established the nature of the individual chemical reactions which make up the polymerization, the authors use a spectroscopic technique to determine rate constants of initiation and propagation.
I.e.:

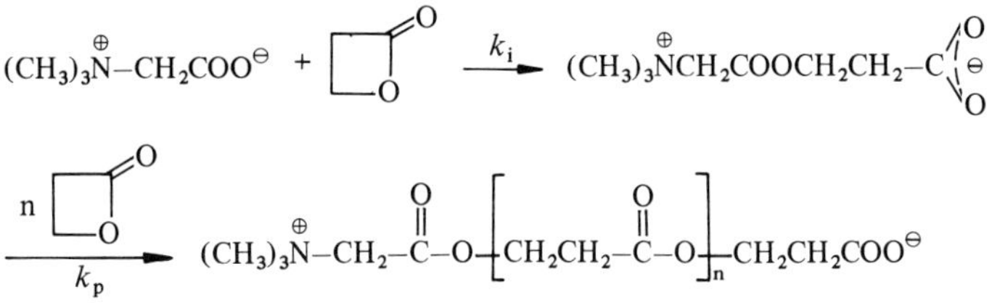

$$(CH_3)_3\overset{\oplus}{N}-CH_2COO^{\ominus} + \underset{\text{(β-propiolactone)}}{\square} \xrightarrow{k_i} (CH_3)_3\overset{\oplus}{N}CH_2COOCH_2CH_2-C\overset{O}{\underset{O}{\diagdown}}{}^{\ominus}$$

$$\xrightarrow[k_p]{n \; \square} (CH_3)_3\overset{\oplus}{N}-CH_2-\overset{O}{\underset{\|}{C}}-O-[CH_2CH_2-\overset{O}{\underset{\|}{C}}-O]_n-CH_2CH_2COO^{\ominus}$$

$k_i = 1.4 \times 10^{-2}$ l · mol^{-1} · s^{-1} at 25 °C

$k_p = 6.2 \times 10^{-2}$ l · mol^{-1} · s^{-1} at 25 °C

The lower value for k_i is attributed to a lower electron density on the betaine carboxylate anion due to the inductive effect of the ammonium group.

It was stated earlier that termination reactions are absent in lactone polymerizations. This is so at solution polymerization temperatures. However, Mayne[45] has shown that during phosphine initiated bulk polymerization at 300 °C the following reaction occurs:

$$\ldots-C\overset{O}{\underset{O}{\diagdown}}{}^{\ominus} + Bu_3\overset{\oplus}{P}-\ldots \longrightarrow \begin{array}{l} \ldots-\overset{O}{\underset{\|}{C}}-OBu + Bu_2P-\ldots \\ \text{or} \\ \ldots-\overset{O}{\underset{\|}{C}}-Bu + Bu_2\overset{O}{\underset{\|}{P}}-\ldots \end{array}$$

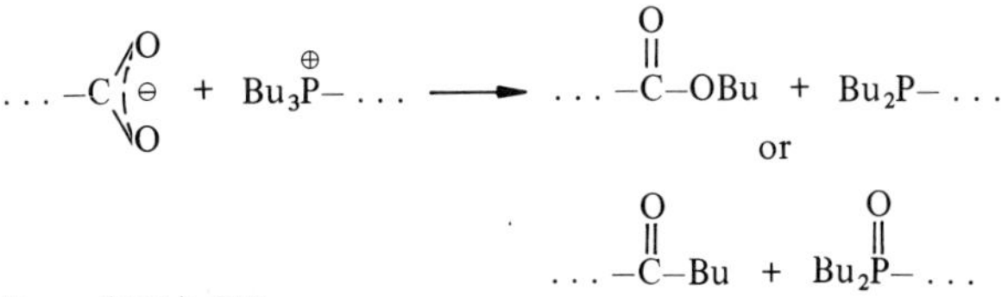

Bu = (CH$_2$)$_3$CH$_3$

Wilson and Beaman[46] found that the polymer formed in the presence of certain cyclic amines contained more than one initiator fragment per chain. They suggest that the following reaction occurs:

$$2 \;\; \overset{\oplus}{\underset{|}{N}}\!\!-\!\!\left[CH_2-\underset{\underset{CH_3}{|}}{\overset{\overset{CH_3}{|}}{C}}-COO\right]_n\!\!CH_2-\underset{\underset{CH_3}{|}}{\overset{\overset{CH_3}{|}}{C}}-C\overset{O}{\underset{O}{\diagdown}}{}^{\ominus}$$

$$\downarrow$$

$$\overset{\oplus}{\underset{|}{N}}\!\!-\!\!\left[CH_2-\underset{\underset{CH_3}{|}}{\overset{\overset{CH_3}{|}}{C}}-COO\right]_n\!\!CH_2-\underset{\underset{CH_3}{|}}{\overset{\overset{CH_3}{|}}{C}}-COOCH_2CH_2-N\!\!-\!\!\left[CH_2-\underset{\underset{CH_3}{|}}{\overset{\overset{CH_3}{|}}{C}}-COO\right]_n\!\!CH_2-\underset{\underset{CH_3}{|}}{\overset{\overset{CH_3}{|}}{C}}-C\overset{O}{\underset{O}{\diagdown}}{}^{\ominus}$$

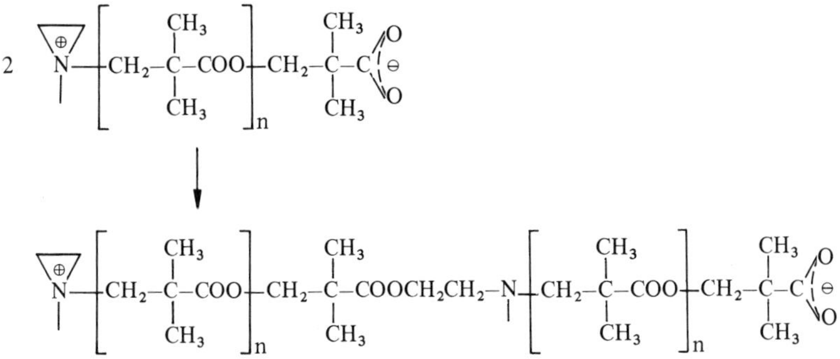

Epoxide/Anhydride

An equimolar mixture of epoxide and cyclic anhydride has the stoichiometry of a polyester. Fisher[47] has shown that if a tertiary amine is added to such a mixture, maintained at 70–100 °C, epoxide and anhydride disappear at identical rates to yield polyester. This relationship held even if the epoxide was present in excess. The excess could be removed at the end of the polymerization and functioned purely as solvent.

When mole ratios of one to one were employed the polymerization order with respect to both monomers was zero up to about 60% conversion. This zero order rate was found to be proportional to the amine concentration and it could be expressed mathematically by the following equation:

$$R_x = a + k \left[\frac{C_x}{C_y}\right] (R_y - a)$$

where R_x and R_y represent the rate at amine concentrations C_x and C_y, respectively. a is the intercept at zero catalyst concentration and k is a constant. Physically a may be taken as the rate of the uncatalysed reaction and k as an efficiency factor representing the portion of the amine that is catalytically effective. For the system studied in detail, allyl glycidyl epoxide and phthalic anhydride at 100 °C, Fisher found a to be 0.07% reaction per minute and k = 96%, i.e., doubling amine increased rate by the factor 1.92.

The following scheme is proposed to explain these observations.

Zwitterion formation:

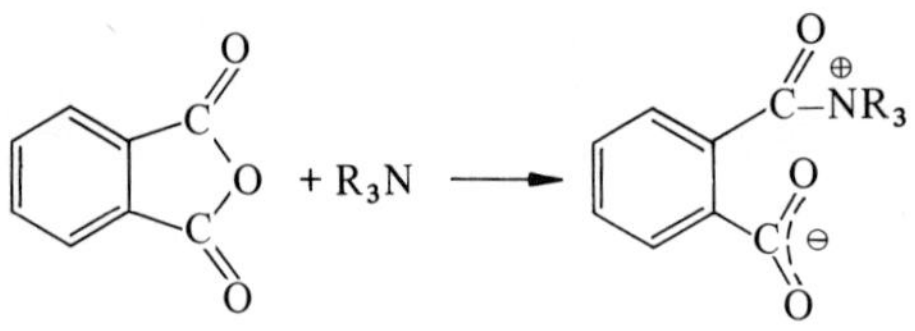

Reaction of carboxylate anion with epoxide:

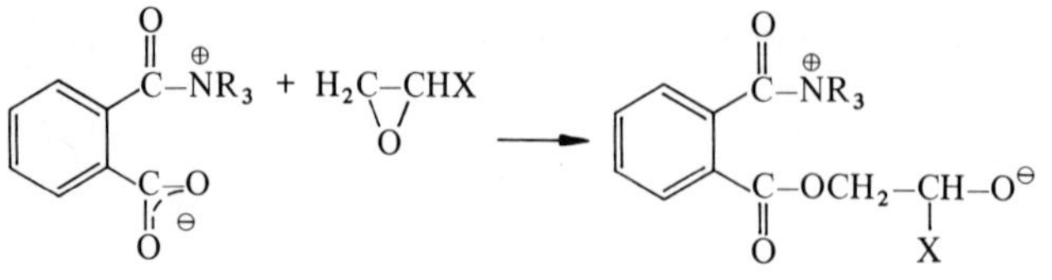

Reaction of alkoxide anion with anhydride:

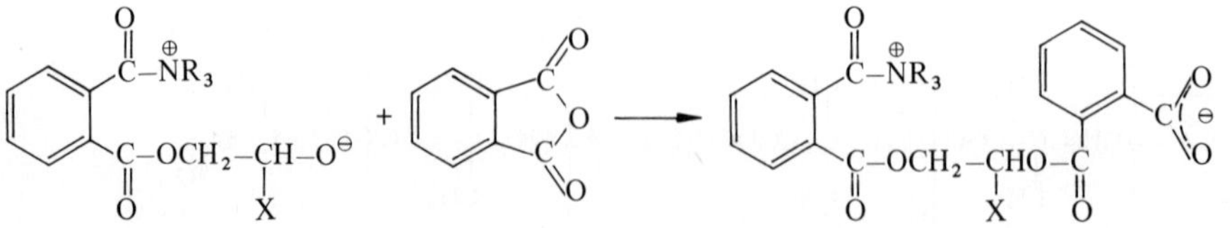

Chain coupling:

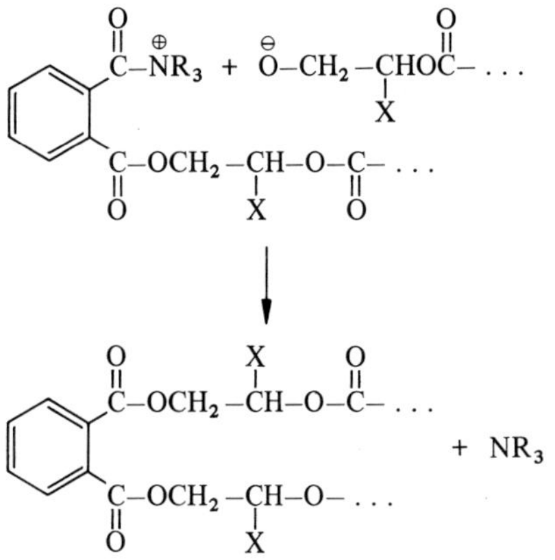

$$\text{Ar-C(=O)-NR}_3^{\oplus} + {}^{\ominus}\text{O-CH}_2\text{-CHOC(=O)-...} \atop {|\atop X}$$

$$\text{Ar-C(=O)-OCH}_2\text{-CH(X)-O-C(=O)-...}$$

↓

$$\text{Ar-C(=O)-OCH}_2\text{-CH(X)-O-C(=O)-...}$$
$$\text{Ar-C(=O)-OCH}_2\text{-CH(X)-O-...} \quad + \text{ NR}_3$$

The molecular weight of the products was quite low, 12 000, but could be raised by careful purification of the monomers. In a tyical run there were ten chains per amine molecule. This leads the author to suggest that the amine is "quite labile". It is unclear what this statement is meant to imply. Loss of amine in the last step would decrease rather than increase the number of chains per amine. In the absence of any evidence for the presence of zwitterions, the authors mechanism must be considered as speculative.

N-Carboxy anhydrides (NCA's)

Base initiated polymerization of N-carboxy anhydrides is a useful synthetic route to polypeptides.

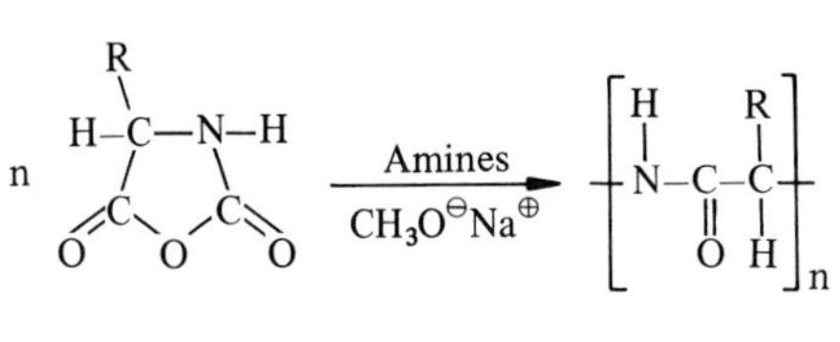

$$n \underset{\text{(NCA)}}{\overset{R}{\underset{|}{\text{H-C-N-H}}}} \xrightarrow[\text{CH}_3\text{O}^{\ominus}\text{Na}^{\oplus}]{\text{Amines}} \left[\text{-N-C-C-} \atop \underset{H}{\overset{H\;\;R}{|\;\;|}} \underset{\;\;\;\;\;O\;H}{\overset{||}{}} \right]_n$$

NCA's have been considered, although the polymerization mechanism appears to be complex and there is no general agreement on the exact nature of the active centre.

Wieland[48] suggested that the growing chains were zwitterions (see page 84).

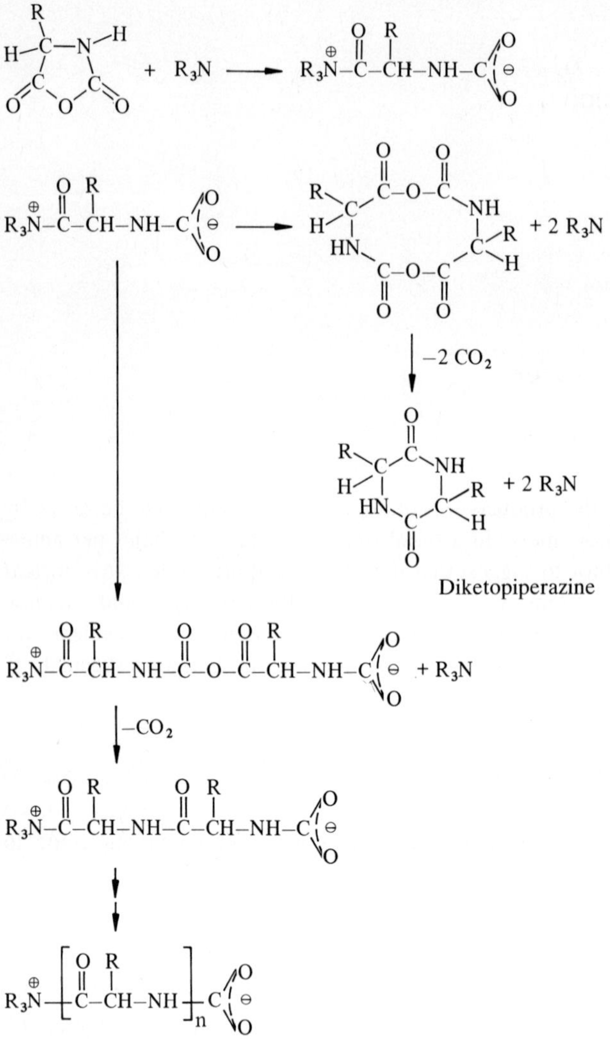

[Reaction scheme showing NCA polymerization mechanism with tertiary amine initiator, formation of diketopiperazine side product, and propagation to polymer]

Kricheldorf and Bosinger[49] compared the polymerization of sarcosine NTA (*J*) and NCA (*K*).

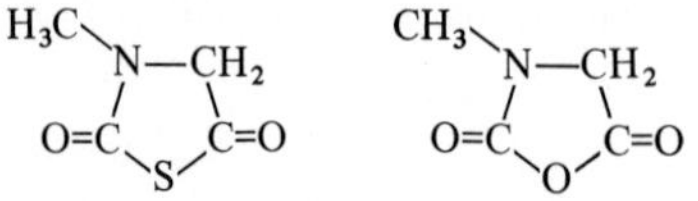

[Structures of J (sarcosine NTA, with S in ring) and K (sarcosine NCA, with O in ring)]

The highest molecular weight polymers were obtained from NCA with pyridine as initiator. The authors put forward a polymerization mechanism similar to that of Wieland. Diketopiperazine was formed during polymerisation of NTA in amounts which decreased as monomer concentration was increased. Increased monomer concentration might be expected to favour growth of a zwitterionic chain relative to its cyclization by intramolecular charge cancellation. Such a phenomenon is well documented in polymerizations where zwitterions definitely are the propagating species.

However, other workers report[50] that *pure* tertiary amines are inactive. Only when small amounts of methanol or water are added does polymerization commence.

1,3,5-Trioxane

Kern and Jaacks[51] investigated the cationic polymerization of 1,3,5-trioxane by boron trifluoride in dichloromethane at 30 °C. Unlike cationic vinyl polymerizations initiated by metal halides, traces of water rather than increasing the polymerization rate, slightly reduced it. Boron trifluoride induced ionisation of water is thereby ruled out as a step in the initiation mechanism.

$$BF_3 + HOR \rightarrow H^{\oplus} + {}^{\ominus}BF_3OR$$

$$H^{\oplus} + Trioxane \rightarrow Polymer$$

In the absence of cocatalysis zwitterion formation is a reasonable postulate.

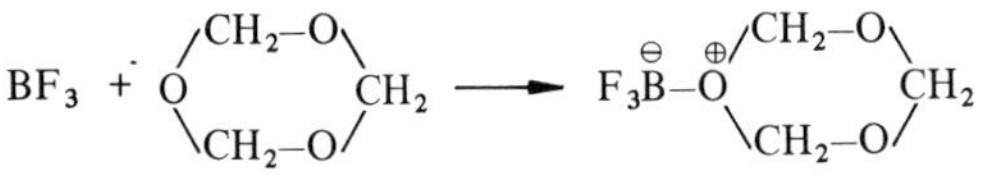

$$BF_3 + \underset{\substack{\diagdown \\ CH_2-O}}{\overset{\diagup CH_2-O\diagdown}{O}} CH_2 \longrightarrow F_3\overset{\ominus}{B}-\overset{\oplus}{\underset{\substack{\diagdown \\ CH_2-O}}{O}}\overset{\diagup CH_2-O\diagdown}{} CH_2$$

$$\longrightarrow F_3\overset{\ominus}{B}-OCH_2OCH_2\overset{\oplus}{O}CH_2 \longleftrightarrow F_3\overset{\ominus}{B}-OCH_2CH_2\overset{\oplus}{O}=CH_2$$

A long induction period is observed and this could be correlated with an increasing formaldehyde concentration. When the concentration of formaldehyde reached a limit (the limit being solely dependent on temperature) polymerization commenced. The onset of polymerization was accelerated by adding formaldehyde.

In all polymerizations an equilibrium exists between active chain end and monomer. This is not usually noticeable because the equilibrium is overwhelmingly in favour of the chain end, and hence, polymer formation. However, the poly(oxymethylene) cation appears to be in equilibrium with a significant concentration of formaldehyde. I.e.:

$$F_3\overset{\ominus}{B}-OCH_2OCH_2\overset{\oplus}{O}CH_2 \rightleftharpoons F_3\overset{\ominus}{B}-OCH_2\overset{\oplus}{O}CH_2 + H_2C=O$$

Monomeric units split off from the chain as formaldehyde. 1,3,5-Trioxane is of course simply the cyclic trimer of formaldehyde.

When BF_3 is added to 1,3,5-trioxane, the rate at which formaldehyde is split off the poly(oxymethylene) chain seems to be much higher than the rate at which new chains are formed. At the start of polymerization when the formaldehyde concentration is low, chain growth is not possible; hence, the induction period.

However, once polymerization commences the rate steadily accelerates. Using an analytical technique the authors were able to show that the growing end population increased with conversion.

The BF$_3$/1,3,5-trioxane system is one of the few so far discovered in which there is a possibility that monomeric units add at the cationic end of a macrozwitterion. Fortunately, the cation seems to be stable in the presence of its counter anion. As a simple model system with which to study cationic propagation through zwitterion intermediates it is marred by its equilibrium nature and the insolubility of the polymer. Whilst kinetic termination seems to be absent, the authors report transfer to the solvent methylene chloride. Such a reaction would introduce non-zwitterionic chains.

2.2 Charge Cancellation Polymerization

Two types of polymerization can be distinguished in this category, those which require other molecules or ions as initiators and those which do not. Pure cyclic sulfonium zwitterions polymerize when heated. The literature gives no indication that diazoalkanes or nitrile oxides will "spontaneously" polymerize in a similar manner.

As mentioned earlier Swarc originally drew attention to the possibility of macrozwitterion formation by charge cancellation coupling of smaller bipolar species.

There are some similarities between this type of polymerization and the coupling of aziridine terminated pivalolactone chains described by Wilson and Beaman[46].

2.2.1 Aryl substituted cyclic Sulfonium Zwitterions

Schmidt et al.[52] have succeeded in synthesizing "monomeric" zwitterions which polymerize when heated.

$$n\ (\overset{\frown}{CH_2})_x\ S^{\oplus}-\!\!\!\left\langle\ \right\rangle\!\!\!-O^{\ominus}\ \longrightarrow\ \left[(CH_2)_x-S-\!\!\!\left\langle\ \right\rangle\!\!\!-O\right]_n$$

The polymerization of two aryl sulfonium zwitterions has been studied in detail.

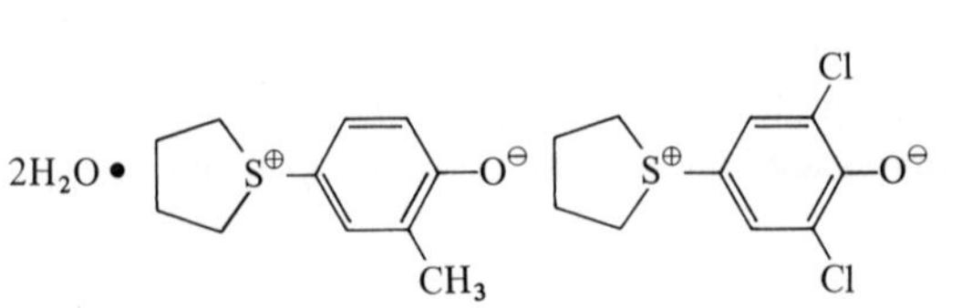

$$2H_2O\bullet\quad L\qquad\qquad M$$

Compound M is much more stable than L. Any attempt to remove water of hydration from L resulted in its rapid polymerization. Stabilisation of crystalline solid hydrated L is probably due to a high interaction energy between zwitterion and water molecules in the crystal lattice.

The increased stability of M suggests that the decrease in nucleophilicity of the phenoxide oxygen caused by the chlorine substituents far outweighs the labilising effect on the perhydrothiophenium ring. Polymerization of M, which can be dried, begins if it is heated above 150 °C. The relative stability of these "monomers" is probably due to a large contribution from the quinoid form to the resonance hybrid.

Nucleophilic attack was only observed at the carbon in the perhydrothiophenium ring adjacent to the sulfur atom. No sulfur-phenyl bond cleavage occurred consistent with a strong π-bond between the sulfonium sulfur and the aromatic system. The dimeric zwitterion formed by the initiation step can react at either end with further "monomer" or other mers of any DP. In fact, the lability of perhydrothiophenium rings on the ends of macrozwitterions would be expected to be greater because of the loss of the adjacent phenoxide anion.

The authors suggest that in a charge cancellation polymerization, propagation will be assisted by zwitterion association. Association of monomeric units has been advanced as an explanation for abnormally fast stereoregular polymerization of acrylic acid in bulk or polar solvents[53].

The molecular weight distributions of polymers obtained by various means are described below:

Table 4. Influence of polymerization technique on polymer molecular weight distribution Monomer L:

Polymerization technique (No.)	Yield in %[a]			
	polymer	High mol. weight cyclics	cyclic trimer	cyclic dimer
Heat. 35 min at 100 °C (a)	80	3	10	7
High vacuum 3 days (b)	80	3	10	7
Boil. 1 h in water (c)	72	3	10	15
Boil. in chlorobenzene (d)	17	12	41	30
Add. NaOCH$_3$ or R$_3$N (e) 0.01 – 0.001 mole-%	80	3	10	7

[a] Determined by GPC

Mass spectrometric analysis of L (technique (a)) gave results consistent with the gel permeation chromatogram, with m/e peaks at 388 (dimer) and 582 (trimer). Pure dimer could be isolated by heating polymer L (technique (a)) and collecting the crystalline sublimate. The molecular weight distributions were quite insensitive to changes in temperature or heating times.

Considerably higher temperatures were required to polymerize M.

Table 5. Influence of polymerization technique on polymer molecular weight distribution Monomer M:

Polymerization technique (No.)	Yield in %[a]			
	polymer	High mol. weight cyclics	cyclic trimer	cyclic dimer
Heat. 170 °C (a)	30	–	–	70
0.01 mole-% R$_3$N (b)	94	–	–	6

[a] Determined by GPC

Polymers obtained from *M* generally had higher molecular weights than those from *L*. When quaternary ammonium ions were inadvertently admixed with monomer, polymers of even higher molecular weight were formed.

The cyclic structure of oligomers was deduced from study of their X-ray diffraction patterns.

From these observations some aspects of chain growth peculiar to zwitterionic propagating species can be deduced. It seems that the proportion of cyclics to long zwitterionic chains is strongly influenced by the dielectric properties of the medium. The most logical explanation is a shift in the inter-intramolecular ion pair equilibrium constant.

The position of minimum free energy will be determined by the balance between chain end electrostatic attraction and chain entropy. An increase in the dielectric constant of the medium will lead to a fall in chain end attraction, an increase in the average end to end distance and a rise in entropy. Fewer macrocycles will be formed.

A similar type of explanation could be proposed for the higher molecular weight of polymer formed in the presence of quaternary ammonium salt. Ion pairing can take place without entropically unfavourable chain cyclization.

The difference in behaviour of the two monomers in the presence of nucleophiles is perhaps to be found in the greater rate of polymerization of *L*. The authors speculate that even at low temperatures many initiations rapidly occur by monomer to monomer attack. The polymerizing mixture at any time contains little of the by comparison less polarized monomer and is unaffected by added nucleophiles.

Where the chains growing from added nucleophiles do form a significant proportion of the polymer (with *M* apparently) higher molecular weight polymers will be formed. Cyclization is no longer possible.

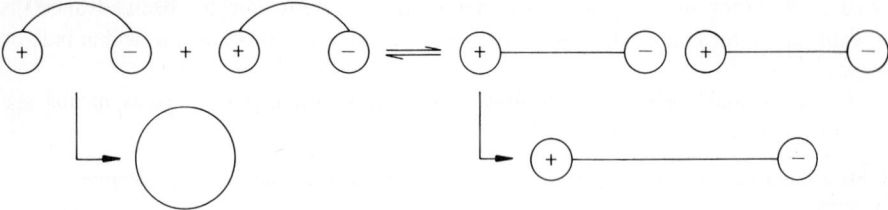

A greater degree of certainty exists with zwitterion monomers that the growing chains are also zwitterions. Hence, they seem to be the most suitable class of compounds with which to study ionic reactions of this type. The concentration of growing zwitterions is high and any peculiarities associated with their growth are likely to be readily apparent. A closer examination of the polymerization of these monomers is warranted.

Schmidt[54] has also carried out a limited amount of work in homogeneous solution. The solvent, a mixture of dipropylene glycol (2%), diethylene glycol (13%) and chlorobenzene (85%), solubilizes both monomer and polymer.

Polymer molecular weights were lower than those obtained by bulk polymerization. This might have been expected, rapid initiation and favouring of cyclization over chain extension would be consequences of the low dielectric constant of the solvent. As well as energetic considerations, the rate of chain extension, a bimolecular process, will be influenced more by reduced monomer concentration than unimolecular cyclization. Despite an increased rate of formation of cyclics, a termination reaction here, the lower stability of zwitterions in this solvent results in faster polymerization rates.

The average molecular weight of the polymer could be increased by further additions of monomer.

2.2.2 Acetonitrile Oxide

Acetonitrile oxide, $H_3C-C\equiv\overset{\oplus}{N}-O^\ominus \leftrightarrow H_3C-\overset{\oplus}{C}=N-O^\ominus$ is bipolar.

Its polymerization has been investigated by Brandi, De Sarlo, and Guarna[55]. Both acetonitrile oxide and benzonitrile oxide form polymers in ethanol. However, the greater solubility of acetonitrile oxide allows reactions with nucleophiles to be carried out at concentrations necessary for the formation of high polymer.

The authors found that the character of the polymers depends strongly on the concentrations of nitrile oxide and nucleophile.

Table 6. Influence of the concentrations of monomer and initiator (nucleophile) on the formation of different products

Nucleophile	Concentrations in mol · l^{-1}		Products
	Nitrile oxide	Nucleophile	
	0.1	0.2	Only 3,4-dimethylfurazan N-oxide
Trimethylamine	1.0	1.0	Predominantly cyclic hexamer, heptamer, and octamer
	3.0	1.0–2.5	Predominantly insoluble high molecular weight polymer
	0.1	0.2	Cyclic dimer
Pyridine	1.0	1.0	Cyclic dimer and hexamer
	3.0	1.0–2.5	Cyclic dimer, hexamer, heptamer, and octamer

In all polymerizations the dimer 3,4 dimethylfurazan N-oxide was a by-product.

Proton and ^{13}C NMR spectra of the oligomers showed that all monomeric units were in identical positions. This is strong indication that the oligomers are not chains. The cyclic structure was confirmed by X-ray crystallography, with an *anti* configuration of the methyl group.

None of the polymers absorbed UV radiation at wavelengths higher than 220 nm. This rules out conjugated double bonds and indicates monomeric units are linked by O–C bonds. Absorptions from C=N and N–O–C linkages were present in IR/Raman spectra. The hexamer has, therefore, the cyclic structure *N*.

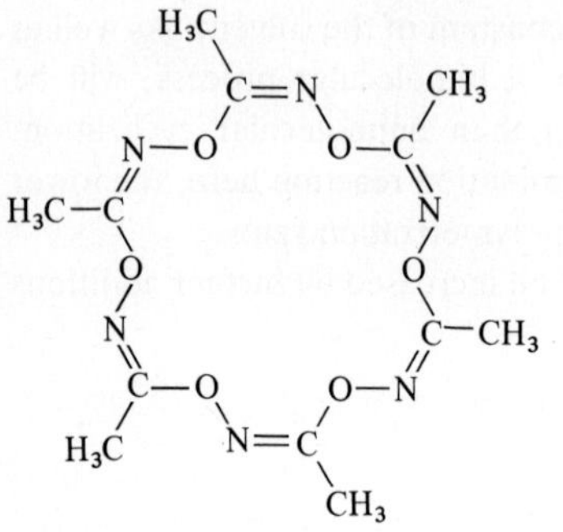

N

The insoluble polymer has the same vibrational spectra as the other polymers but breaks down upon heating and explodes under laser irradiation. It is believed to have the linear structure *O*.

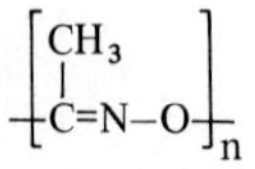

O

The authors put forward the following polymerization mechanism:

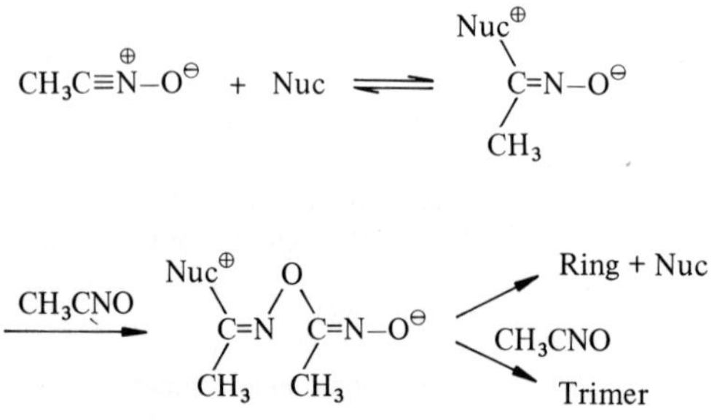

$$CH_3C\overset{\oplus}{\equiv}N-O^{\ominus} + Nuc \rightleftharpoons \underset{CH_3}{\overset{Nuc^{\oplus}}{C}}=N-O^{\ominus}$$

$$\xrightarrow{CH_3CNO} \underset{CH_3\ CH_3}{\overset{Nuc^{\oplus}\ O}{C=N\ C}}=N-O^{\ominus} \quad \begin{matrix} \rightarrow Ring + Nuc \\ \leftarrow CH_3CNO \\ \searrow Trimer \end{matrix}$$

Cyclization by intramolecular displacement of amine and simple monomer addition are in competition. Increase in monomer concentration will not affect the rate of cyclization but will increase the rate of chain growth. Hence, the dependence of average molecular weight on monomer concentration.

De Sarlo et al.[55] suggest that the stereochemical course of the polymerization leading exclusively to cyclic polymers with methyl groups in anti-positions is in agreement with the usual *trans*-addition to unsaturated bonds. The nucleophilic centre binds to the carbon atom of the nitrile oxide triple bond, and the lone pair settles on the nitrogen atom in *trans*-orientation. This behaviour was also observed when hydrogen chloride added to nitrile oxides.

However, an alternate explanation could be advanced to explain the stereochemistry. At low monomer concentrations, which favour cyclization, ion pairing is most likely intramolecular.

trans-Addition:

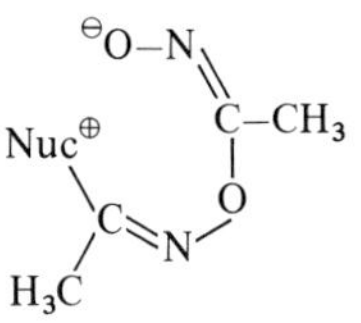

$^\ominus$O—N
          \\
           C—CH$_3$
          /
       O
        \\
         C=N
        /
     H$_3$C

Nuc$^\oplus$

cis-Addition:

H$_3$C
     \\
      C=N
          \\
           O
            \\
             C=N—CH$_3$
            /
         $^\ominus$O

Nuc$^\oplus$

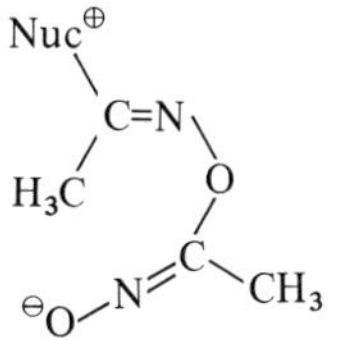

trans-Addition leads to cyclic structures in which the oppositely charged chain ends are closer together.

Monomer molecules can insert between the ion pair *only* in the *trans*-mode. It should be noted that propagation takes place under these circumstances without charge separation.

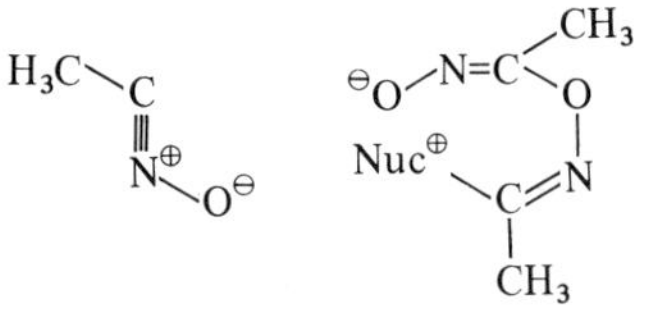

H$_3$C—C
       ‖
       N$^\oplus$
        \\
         O$^\ominus$

$^\ominus$O—N=C
              \\
               CH$_3$
              /
           O
            \\
             C≡N
            /
         Nuc$^\oplus$
              |
              CH$_3$

Stereoregular chain growth of this type could be a consequence of the zwitterionic nature of the chains.

One mode of chain growth, zwitterion interaction, has not been considered by the authors.

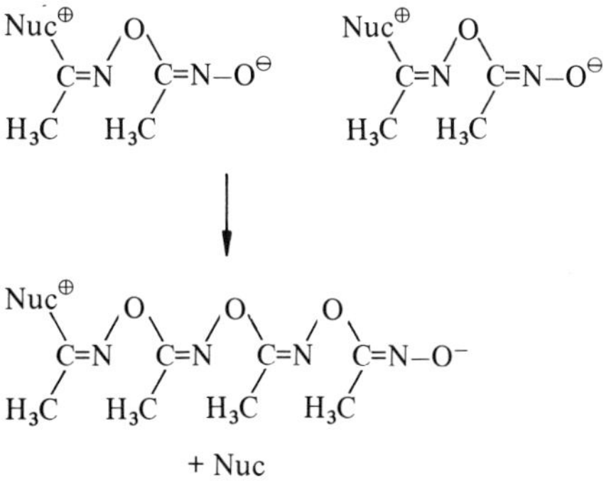

Nuc$^\oplus$ O
         \\ / \\
          C=N   C=N—O$^\ominus$
         /   /
      H$_3$C H$_3$C

Nuc$^\oplus$ O
         \\ / \\
          C=N   C=N—O$^\ominus$
         /   /
      H$_3$C H$_3$C

↓

Nuc$^\oplus$ O   O
         \\ / \\ / \\
          C=N   C=N   C=N   C=N—O$^-$
         /   /   /   /
      H$_3$C H$_3$C H$_3$C H$_3$C

+ Nuc

This might be a more rational explanation for the chain-ring ratio increasing with *both* initiator and monomer concentrations.

2.2.3 Diazoalkanes

Alkyl borates, boron alkyls, and boron halides have all been used to polymerize diazoalkanes. Several authors have claimed that macrozwitterions are formed[56-58].

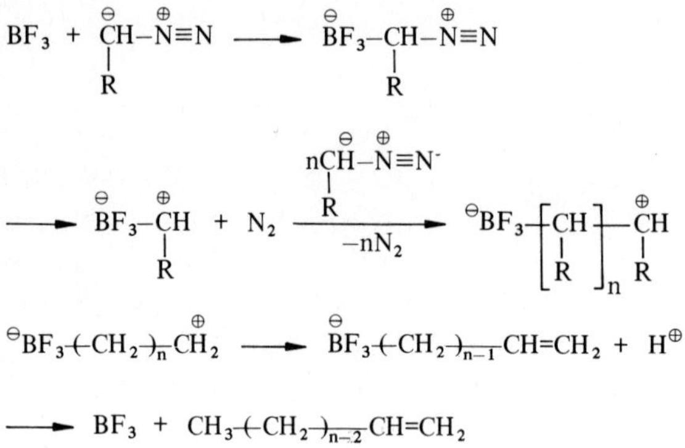

$$BF_3 + \overset{\ominus}{C}H-\overset{\oplus}{N}\equiv N \longrightarrow \overset{\ominus}{B}F_3-\underset{R}{C}H-\overset{\oplus}{N}\equiv N$$

$$\longrightarrow \overset{\ominus}{B}F_3-\underset{R}{C}H + N_2 \xrightarrow[-nN_2]{n\overset{\ominus}{C}H-\overset{\oplus}{N}\equiv N^-} \overset{\ominus}{B}F_3-\left[\underset{R}{CH}\right]_n-\underset{R}{\overset{\oplus}{C}H}$$

$$\overset{\ominus}{B}F_3\text{-}(CH_2)_n\text{-}\overset{\oplus}{C}H_2 \longrightarrow \overset{\ominus}{B}F_3\text{-}(CH_2)_{n-1}\text{-}CH=CH_2 + H^{\oplus}$$

$$\longrightarrow BF_3 + CH_3\text{-}(CH_2)_{n-2}\text{-}CH=CH_2$$

Two objections have been raised to this mechanism, first that charge separation would be energetically unfeasible[59] and second, that the very reactive carbonium ion would rearrange[60].

It seems that co-catalysis can be ruled out, alcohols do not accelerate the rate of polymerization[59].

Bawn, Ledwith, and Matthies[59] have put forward a polymerization scheme which is consistent with all the known facts. They suggest that, depending on the stage of polymerization, fluoride or $-(CH_2)_nF$ anions migrate to and neutralize the cation.

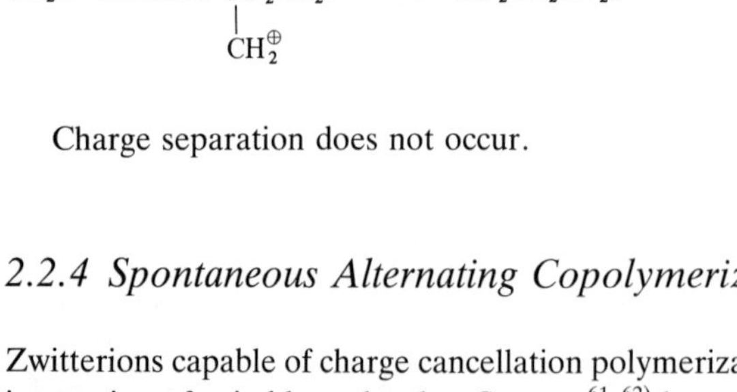

$$F^{\ominus}\text{-Transfer:} \quad \overset{\ominus}{B}F_3\text{-}\overset{\oplus}{C}H_2 \longrightarrow BF_2\text{-}CH_2F$$

$$CH_2N_2 + BF_2CH_2F \longrightarrow \underset{\overset{\oplus}{C}H_2N_2}{\overset{\ominus}{B}F_2CH_2F} \longrightarrow \underset{\overset{\oplus}{C}H_2}{\overset{\ominus}{B}F_2CH_2F}$$

$$CH_2F\text{-Transfer:} \quad \underset{\overset{\oplus}{C}H_2}{\overset{\ominus}{B}F_2CH_2F} \longrightarrow BF_2CH_2CH_2F$$

Charge separation does not occur.

2.2.4 Spontaneous Alternating Copolymerizations

Zwitterions capable of charge cancellation polymerization can be formed "in situ" by the interaction of suitable molecules. Saegusa[61, 62] has used this method to produce a wide range of alternating copolymers. For instance a quantitative yield of alternating copolymer is obtained when equimolar amounts of 2-oxazoline and β-propiolactone are mixed in an aprotic solvent at room temperature.

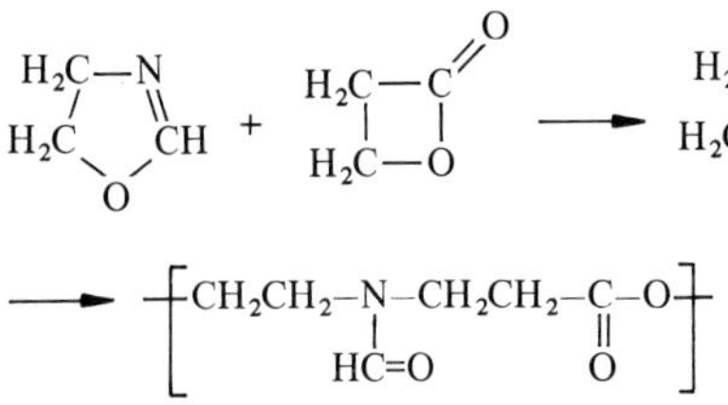

Macrozwitterion Polymerization

[Scheme: oxazoline + β-propiolactone reaction]

$$\text{H}_2\text{C}-\text{N} \atop \text{H}_2\text{C} \quad \text{CH} \atop \text{O} \ \ \searrow \!\!\!\!\nearrow + \begin{matrix}\text{H}_2\text{C}-\text{C}{=}\text{O}\\ | \quad |\\ \text{H}_2\text{C}-\text{O}\end{matrix} \longrightarrow \text{H}_2\text{C}-\overset{\oplus}{\text{N}}-\text{CH}_2\text{CH}_2\text{CO}_2^{\ominus}$$

$$\longrightarrow \left[\text{CH}_2\text{CH}_2-\underset{\underset{\text{HC=O}}{|}}{\text{N}}-\text{CH}_2\text{CH}_2-\overset{\text{O}}{\overset{\|}{\text{C}}}-\text{O} \right]$$

In this example oxazoline functions as nucleophile and β-propiolactone as electrophile.

Molecular weight/conversion data show that the number of copolymer molecules rises initially, reaches a maximum at 80% conversion and then steadily falls. The implication is that zwitterions form more rapidly than they combine. If the oxazoline mole ratio is below one half, the polymer contains less than 50% oxazoline units. Under these conditions k_p [Zwitterion]^2 and k[Zwitterion] [β-Propiolactone] are comparable.

Other monomers which have been shown to undergo this type of polymerization are listed below (see Table 7 on page 94).

Macrocycles were not detected among any of the products. However, cyclization of the initial zwitterion did occur with the two dioxaphospholane nucleophiles and several electrophiles[63].

E.g.:

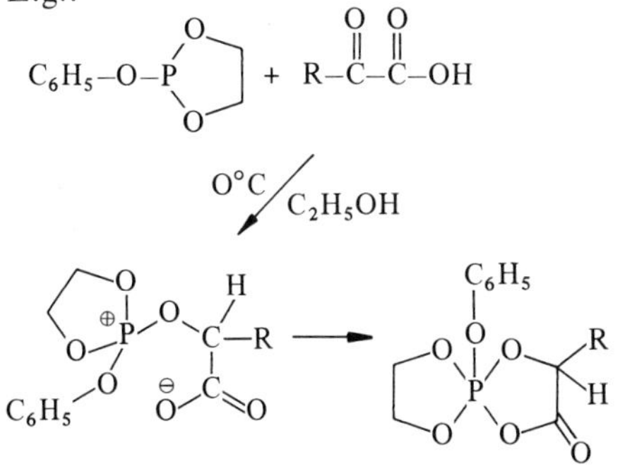

[Scheme: dioxaphospholane + R-CO-COOH → cyclic phosphorane product, with 0°C / C$_2$H$_5$OH]

Ether, a solvent with a low dielectric constant, favours cyclization over chain extension. Heating the phosphorane to 120 °C yields the alternating copolymer. Terpolymers are formed when either acrylates or acrylonitrile are copolymerized with dioxaphospholane in a carbon dioxide atmosphere.

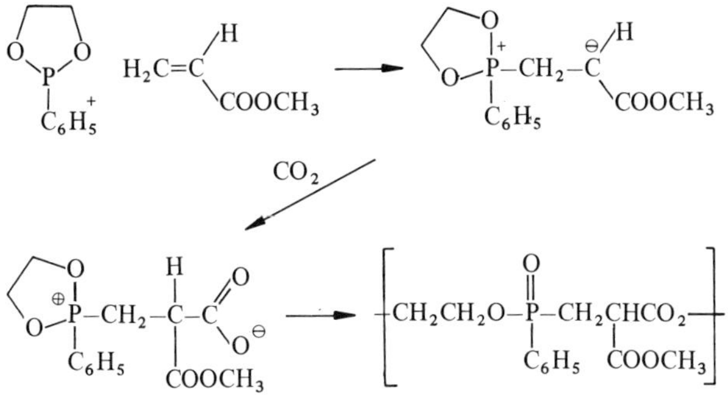

[Scheme: dioxaphospholane + H$_2$C=CH(COOCH$_3$) → zwitterion intermediate → with CO$_2$ → terpolymer $[\text{CH}_2\text{CH}_2\text{O}-\text{P}(\text{C}_6\text{H}_5)(=\text{O})-\text{CH}_2\text{CHCO}_2]$ with COOCH$_3$ side group]

Table 7. Monomer combinations for alternating copolymerizations
M_N: nucleophilic monomer; M_E: electrophilic monomer. $M_N + M_E \rightarrow M_N^\oplus - M_E^\ominus$

M_N	$M_N^\oplus$	M_E	$-M_E^\ominus$
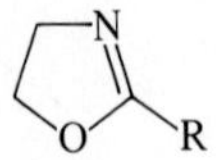 oxazoline (R)	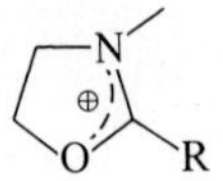 oxazolinium (R)	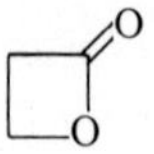 β-propiolactone	$-CH_2CH_2CO_2^\ominus$
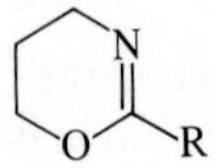 5,6-dihydro-oxazine (R)	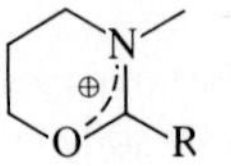 5,6-dihydro-oxazinium (R)	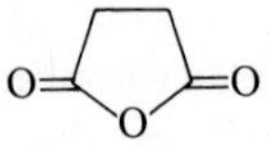 succinic anhydride	$-\overset{O}{\underset{\|}{C}}-CH_2CH_2CO_2^\ominus$
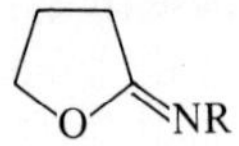 2-oxazoline (NR)	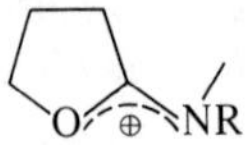 2-oxazolinium (NR)	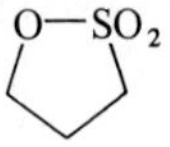 1,3-propanesultone (O–SO$_2$)	$-(CH_2)_3-SO_3^\ominus$
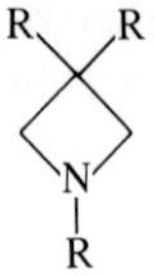 azetidine (R,R,R)	 azetidinium (R,R,R)	$H_2C=C\overset{H}{\underset{COOH}{}}$	$-CH_2CH_2-CO_2^\ominus$
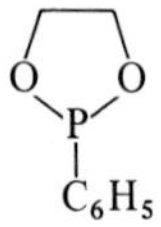 dioxaphospholane (C_6H_5)	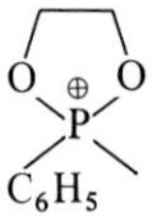 dioxaphospholanium (C_6H_5)	$H_2C=C\overset{H}{\underset{CONH_2}{}}$; 2-pyrrolidinone 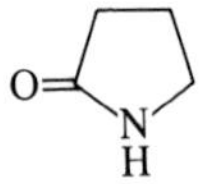	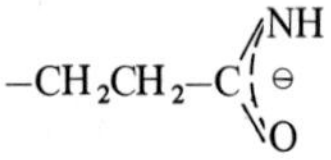 $-CH_2CH_2-C\overset{NH}{\underset{O}{\|}}{}^\ominus$
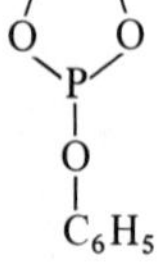	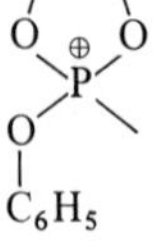	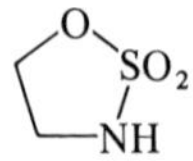 cyclic sulfamate (O–SO$_2$, NH)	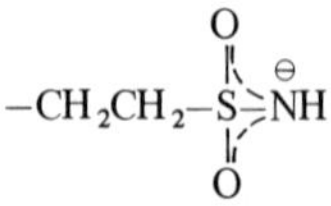 $-CH_2CH_2-\overset{O}{\underset{O}{S}}{\geq}NH^\ominus$
dioxaphosphorinane (C_6H_5)	dioxaphosphorinanium (C_6H_5)	$H_2C=C\overset{H}{\underset{COR}{}}$ R = CH$_3$; OCH$_3$	$-CH_2CH-$ | COOR
		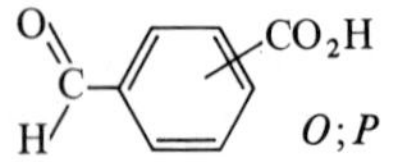 $\underset{H}{\overset{O}{C}}{\diagdown}\text{–C}_6H_4\text{–}CO_2H$ O;P	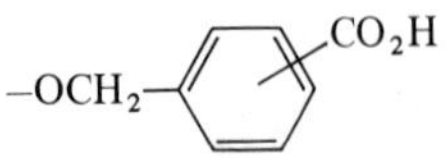 $-OCH_2-\text{C}_6H_4\text{–}CO_2H$
		$R-\underset{\underset{O}{\|\|}}{C}-\underset{\underset{O}{\|\|}}{C}-OH$	$-O-\underset{R}{\overset{\|}{CH}}-\underset{O}{\overset{\|\|}{C}}-O^\ominus$
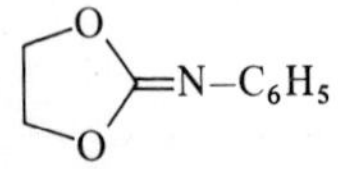 dioxolane (=N–C$_6H_5$)	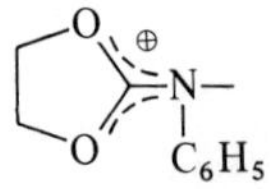 dioxolanium (N–C$_6H_5$)	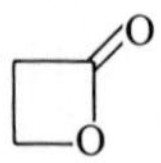 β-propiolactone	$-CH_2CH_2-CO_2^\ominus$

To date Saegusa seems to have forsaken detailed mechanistic studies on any system to demonstrate the synthetic utility of his so called "spontaneous alternating copolymerizations".

He lists three other polymerizations with similarities to those he describes:

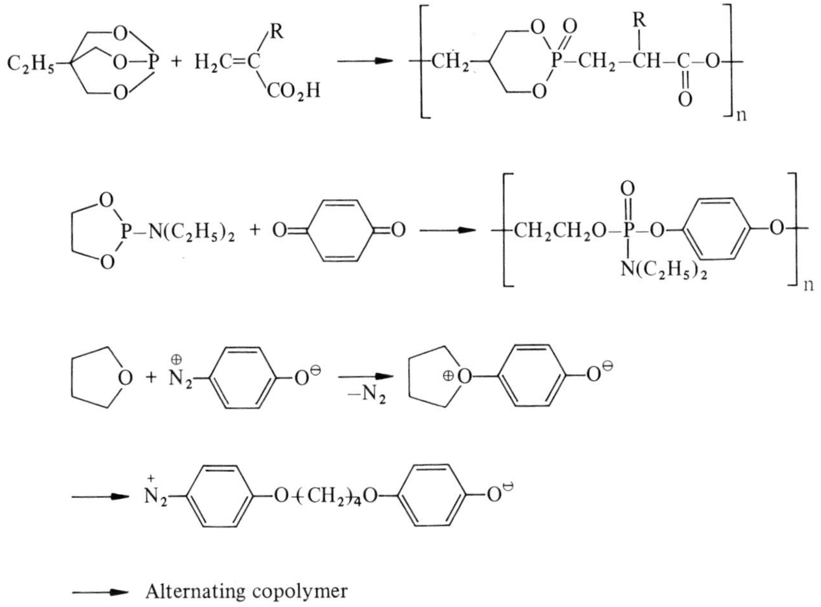

[Reaction scheme 1: C₂H₅-cyclic(O-P(O)-O) + H₂C=C(R)(CO₂H) → [-CH₂-cyclic(O-P(=O)-O)-CH₂-CH(R)-C(=O)-O-]ₙ]

[Reaction scheme 2: cyclic(O-P-N(C₂H₅)₂-O) + O=⟨benzene⟩=O → [-CH₂CH₂O-P(=O)(N(C₂H₅)₂)-O-⟨benzene⟩-O-]ₙ]

[Reaction scheme 3: tetrahydrofuran + N₂⁺-⟨benzene⟩-O⁻ → (−N₂) → tetrahydrofuran-O-⟨benzene⟩-O⁻ → N₂⁺-⟨benzene⟩-O-(CH₂)₄-O-⟨benzene⟩-O⁻ → Alternating copolymer]

3 Mechanism of Macrozwitterion Polymerization

In this section an attempt is made to construct a general scheme into which as many as possible of the conclusions reached about macrozwitterion polymerization can be fitted. It has already been pointed out that chemical reactions between electrophilic and nucleophilic molecules which yield charged products or intermediates have been studied by organic chemists for many years. For a polymerization to occur, a reaction pathway for consecutive addition of at least one molecule to the bipolar species must be available. Such reactions are properly the domain of the polymer chemist. Formation of high molecular weight polymer requires that charges be separated until there is no longer any inductive or electrostatic interaction between them. Various authors have realized that this cannot be accomplished with the enthalpy, released when monomer bonds are broken.

This fact has prompted four suggestions:
(i) Macrozwitterion polymerizations do not occur[59, 64, 65]
(ii) Polymerization involves insertion of monomer between inter or intramolecular ion pairs[9, 12, 43].
(iii) Interactions between the poles of the zwitterion are reduced by solvation[43].
(iv) The initial polymerization steps are reversible[35, 37, 51].

There is now ample evidence available to show that macrozwitterion polymerizations do occur and that steps (ii)–(iv) play a role in charge separation. The following general mechanism is, therefore, indicated:

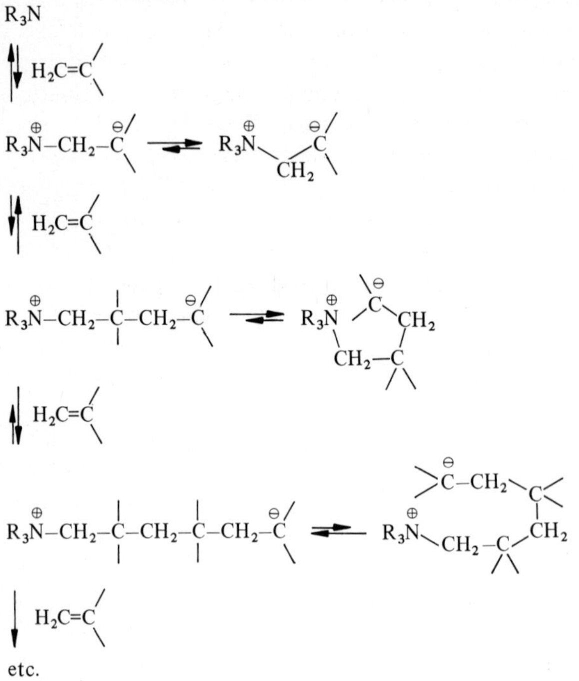

$$R_3N$$

$$\downarrow H_2C=C\begin{smallmatrix}/\\\\\backslash\end{smallmatrix}$$

$$R_3\overset{\oplus}{N}-CH_2-\overset{\ominus}{C}\begin{smallmatrix}/\\\backslash\end{smallmatrix} \rightleftharpoons R_3\overset{\oplus}{N}\diagdown\begin{smallmatrix}\\CH_2\end{smallmatrix}\overset{\ominus}{C}\begin{smallmatrix}/\\\backslash\end{smallmatrix}$$

$$\downarrow H_2C=C\begin{smallmatrix}/\\\backslash\end{smallmatrix}$$

$$R_3\overset{\oplus}{N}-CH_2-\overset{|}{C}-CH_2-\overset{\ominus}{C}\begin{smallmatrix}/\\\backslash\end{smallmatrix} \rightleftharpoons R_3\overset{\oplus}{N}\diagdown\begin{smallmatrix}\overset{\ominus}{C}\\CH_2-C\end{smallmatrix}\diagdown CH_2$$

$$\downarrow H_2C=C\begin{smallmatrix}/\\\backslash\end{smallmatrix}$$

$$R_3\overset{\oplus}{N}-CH_2-\overset{|}{C}-CH_2-\overset{|}{C}-CH_2-\overset{\ominus}{C}\begin{smallmatrix}/\\\backslash\end{smallmatrix} \rightleftharpoons R_3\overset{\oplus}{N}\diagdown\begin{smallmatrix}\overset{\ominus}{C}-CH_2\diagdown C\\CH_2-C\end{smallmatrix}\diagdown CH_2$$

$$\downarrow H_2C=C\begin{smallmatrix}/\\\backslash\end{smallmatrix}$$

etc.

The diagram uses a tertiary amine and a vinyl monomer for illustration, but is meant to be entirely general. Obviously the relative importance of each step will depend on the initiator-monomer combination, the solvent, and temperature.

Charge cancellation polymerizations must be considered as special cases. Some degree of charge separation already exists in acetonitrile oxide but a base is required to initiate its polymerization. Although linked by a conjugated system ionisation is essentially complete in cyclic sulfonium zwitterion monomers. No initiator is required and monomeric units add at both ends of the polymer chain. The initial step in the general scheme above is best considered as part of monomer synthesis.

The nature of spontaneous alternating copolymerizations makes it impossible to distinguish an initiator and monomer and reaction occurs at both chain ends.

3.1 Covalent Initiators

Cationic polymerizations thought to involve macrozwitterions are too few in number for any general conclusions to be drawn about initiators. This section will be concerned solely with anionic polymerizations.

The covalent initiators employed are almost always tertiary amines or phosphines. However, Künzel, Giefer, and Kern[43], who studied formaldehyde polymerization, compared triphenylamine, phosphine, stilbene, and arsine. They found a rough correlation between the rate of polymerization and the dipole moment of the initiator. Dipole moments are determined by a number of factors, one of which is bond polarizability. It is well known that the difference in nucleophilicity between nitrogen and phosphorus is mainly due to the higher polarizability of phosphorus. The electron pair is more available for reaction[66]. Inspection of the literature makes it quite clear that nucleophilic phosphorus is invariably the most effective initiator of macrozwitterion polymerization.

Basicity, usually expressed as pK_a, is a poor general indicator of nucleophilicity. Alkylamines, weakly nucleophilic toward electron deficient carbon, have pK_a's eight orders of magnitude higher than triphenylphosphine. However, as a means for assessing nucleophilicity within amines, pK_a's are somewhat more useful. As initiators of both formaldehyde and vinyl polymerizations, aromatic amines (pK_a 5.0) have been found to be much less effective than the lower n-alkylamines (pK_a's 11.0).

There are, however, two exceptions to this rule. (i) Amines with highly branched alkyl substitutents are significantly less active than their straight chain analogues. (ii) Pyridine (pK_a 5.2) has roughly the same initiating power as the far more basic aliphatic amines. These apparent anomalies are evident in both formaldehyde and vinyl polymerizations.

Two possible explanations are suggested when reference is made to the general macrozwitterion polymerization scheme. Quaternization of aliphatic amines will force the three alkyl substituents into a more nearly tetrahedral configuration and increase steric repulsions between them. Enthalpy will have to be expended to overcome this interaction. Charge separation is energetically unfavourable and with the weakly nucleophilic amines even small amounts of enthalpy, lost in this way, may markedly reduce the initiation rate. Clearly, quaternization of pyridine cannot change the orientation of the substituents attached to nitrogen. The aliphatic amine triethylenediamine, whch has its alkyl groups locked in position, has been found to be a very effective initiator of lactone polymerization[44].

Secondly, the nature of the interaction between the monomer-initiator adduct and solvent will be of importance. Co-ordination of a donor solvent to the ammonium ion might be expected to reduce its deactivating inductive influence on the anion. This is analogous to the frequently observed difference in reactivity between free (solvated) and paired ions. Bulky substituents on nitrogen would prevent the formation of a tight solvation shell around it. In this connection two observations, made by Kern et al.[43] on formaldehyde polymerization, are very relevant.
1. Pyridine, while as effective an initiator as alkylamines in the donor solvents acetone and ether, did not cause polymerization in toluene.
2. Despite the similarity of their pK_a's, compared to pyridine, quinoline is a much weaker initiator.

Thus, in non donor solvents, pK_a may be a reasonably accurate reflection of the nucleophilicity of amines with non branched substituents.

3.2 Reactivity of Covalent Base-Monomer Adducts

The investigators of the five polymerizations listed below have produced convincing evidence that the first steps in macrozwitterion polymerization are either reversible or slower than true propagation.

1. Propiolactone	(β-propiolactone structure)	;	Betaine $(CH_3)_3\overset{\oplus}{N}CH_2COO^{\ominus}$	Ref. 43
2. Acrylonitrile	$H_2C=C\overset{H}{\underset{CN}{}}$	;	Triethyl phosphite $(C_2H_5O)_3P$	Ref. 17
3. Cyanoacrylate	$H_2C=C\overset{CN}{\underset{COOR}{}}$	;	Amines $C_6H_5CH_2N(CH_3)_2$; pyridine	Ref. 34, 35
4. Vinyl ether	$H_2C=C\overset{H}{\underset{OR}{}}$	;	DDQ (2,3-dichloro-5,6-dicyano-1,4-benzoquinone)	Ref. 37
5. Trioxane(formaldehyde)		;	BF_3	Ref. 51

The rate of amine initiated cyanoacrylate polymerizations increases as the temperature is reduced. Johnston and Pepper have shown that the initial polymerization steps are reversible. As they are exothermic the concentration of species which initiate the formation of long chain polymer rises as the temperature falls. The fact that initiation is reversible with the very reactive cyanoacrylate monomers suggests that this may be a more common phenomenon than is generally appreciated. Few authors have investigated the effect on rate of temperature change. Grodzinsky et al.[25] have found that the rate of pyridine initiated nitroethylene polymerizations also increases as temperature is reduced. The authors believe that there are transfer-termination reactions which have higher activation energies than initiation or propagation. However, the experimental evidence does not rule out initiation equilibria.

Johnston and Peppers results suggest that the equilibrium betaine concentration is strongly influenced by the nature of the substituents the betaine carries:

$$\underset{/}{\overset{\backslash}{-}}N: + CH_2=C\overset{CN}{\underset{COOR}{|}} \underset{}{\overset{K_1}{\rightleftharpoons}} \underset{/}{\overset{\backslash}{-}}\overset{\oplus}{N}-CH_2-\overset{CN}{\underset{COOR}{\underset{|}{C^{\ominus}}}} + CH_2=C\overset{CN}{\underset{COOR}{|}} \underset{}{\overset{K_2}{\rightleftharpoons}} \underset{/}{\overset{\backslash}{-}}\overset{\oplus}{N}-CH_2-\overset{CN}{\underset{COOR}{\underset{|}{C}}}-CH_2-\overset{CN}{\underset{COOR}{\underset{|}{C^{\ominus}}}}$$

Initiator utilisation during ethyl cyanoacrylate polymerizations differs widely between amines. Only a few per cent of pyridine forms betaine, whereas benzyldimethylamine reacts quantitatively. Despite this at 20 °C pyridine initiated polymerizations are two orders of magnitude faster. Clearly the pyridinium betaine is much the stronger nucleophile. It was suggested in the previous section that the nitrogen atom of pyridine is readily accessible to solvent. Cyanoacrylate polymerizations were conducted in the strong donor solvent tetrahydrofuran. The pyridinium ion is, therefore, probably strongly solvated and its inductive influence on the anion very much reduced. For the aliphatic amine, K_2 is small, K_1 large, for pyridine exactly the reverse.

The electron donating/withdrawing power of substituents at the α-carbon of the monomer also influence the magnitude of K_1 and K_2. This is well illustrated by the contrasting effects of change in size of alkyl group substituents on the rate of cationic and anionic polymerization.

I.e.:

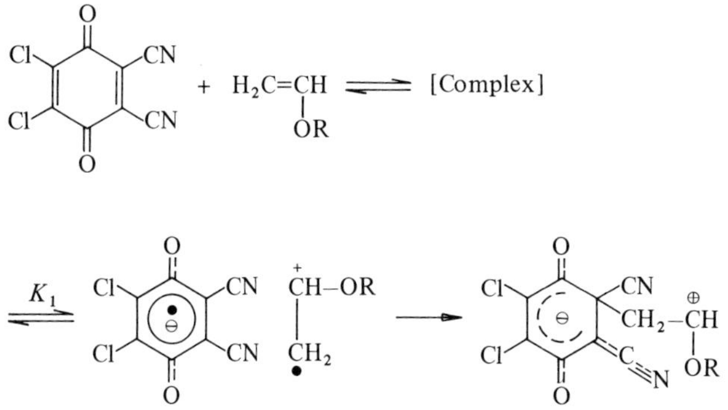

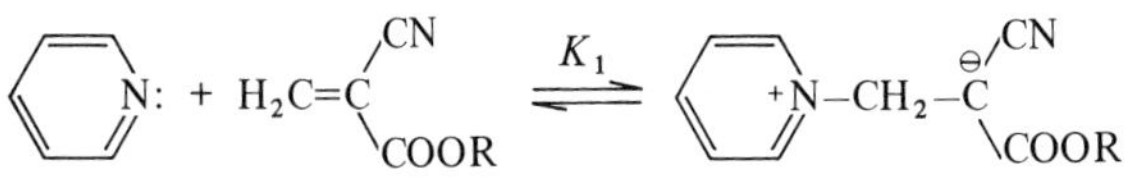

$$\text{Cl-(quinone with CN, CN, Cl)} + H_2C=CH\text{-OR} \rightleftharpoons [\text{Complex}]$$

$$\xrightleftharpoons{K_1} \text{Cl-(quinone radical anion with CN, CN, Cl)} \overset{+}{C}H\text{-OR} / \overset{\bullet}{C}H_2 \longrightarrow \text{Cl-(quinone with CN, CN=C-OR, Cl)}CH_2-\overset{\oplus}{C}H$$

$$\text{Py-N:} + H_2C=C\overset{CN}{\underset{COOR}{}} \xrightleftharpoons{K_1} \text{Py-}\overset{+}{N}-CH_2-\overset{\ominus}{C}\overset{CN}{\underset{COOR}{}}$$

When a series of monomers is compared it is found that in the first reaction K_1 is larger, in the second *smaller*, the larger the alkyl group.

This effect has an interesting consequence for the reactivity of the cyanoacrylate esters. Ethyl betaine will not initiate polymerization of butyl monomer. During addition of BCA to an ECA betaine not only would the stabilising influence of the adjacent ammonium ion be lost, but a thermodynamically less stable anion would have to be formed.

3.3 The Influence of Chemically Inert Salts on the Polymerization Rate

A number of authors have shown that added inert salt can increase the rate of a macrozwitterion polymerization. The polymerizations concerned are:

$(C_6H_5)_3P-CH_2^\bullet$ LiBr; $H_2C=C\overset{CH_3}{\underset{COOCH_3}{}}$ Ref. 9

$(C_2H_5O)_3P$; $H_2C=C\overset{H}{\underset{CN}{}}$ Ref. 17

R_3N ; $H_2C=C\overset{CN}{\underset{COOR}{}}$ Ref. 35

This is not typical of anionic polymerizations in general. Added inert salt usually decreases the rate of polymerizations by reducing the ratio of free to paired ions. Various explanations have been offered and they can be grouped into three basic ideas.

(i) The monomer-initiator adduct (betaine) is weakly nucleophilic. The anion of the salt co-ordinates to the betaine and makes it more nucleophilic.
(ii) Cyclic ion pairs dissociate and pair with the ions of the salt. Monomer addition is no longer opposed by a fall in chain entropy or the enthalpy required to separate charged chain ends.
(iii) The betaine or even second adduct is formed reversibly. The salt co-ordinates to the bipolar species, stabilizes it and reduces the rate of the back reaction. The equilibrium constant is thereby increased.

The first explanation ignores the counter cation of the salt

$$^{\oplus}X-Y^{\ominus} + A^{\oplus}B^{\ominus} \rightleftharpoons B^{\ominus}X^{\oplus}-Y^{\ominus}A^{\oplus}$$

Only when $Y^{\ominus}A^{\oplus}$ is highly dissociated could the effective nucleophilicity of $Y^{\ominus}$ be increased.

Evidence to support hypothesis (ii) comes from Schmidt's observation that when salt is added, fewer macrocycles are formed during cyclic sulfonium zwitterion polymerizations.

If one of the polymerizations listed above had been shown not to involve pre-initiation equilibria some doubt would have been cast on the last postulate. Johnston and Pepper found that the following equilibria exist during cyanoacrylate polymerization.

$$\text{Py} + \text{CA} \rightleftharpoons {}^{\oplus}\text{PyCA}^{\ominus} \xrightarrow{\text{CA}} {}^{\oplus}\text{PyCA CA}^{\ominus}$$

$$\text{BzMe}_2\text{N} + \text{CA} \rightarrow \text{BzMe}_2\text{N}^{\oplus}\text{CA}^{\ominus} \xrightleftharpoons{\text{CA}} \text{BzMe}_2\text{N}^{\oplus}\text{CA CA}^{\ominus}$$

Salts have a much greater influence on aliphatic amine initiated polymerizations. If the primary effect of salt was to influence the position of equilibria this difference might not have been expected. In the second case there is a significant concentration of betaine present for salt to coordinate to and activate by mechanism (ii).

3.4 The Role of Solvent in Charge Separation

Those authors who have studied zwitterionic polymerizations in more than one solvent find that the more "polar" the solvent the faster the polymerization.

Polarity is a general term which covers two solvent properties, solvating power and dielectric constant. Neutral species ionize more readily in strongly co-ordinating solvents. Once ionization has taken place the degree to which the ions dissociate is determined by the solvent dielectric constant. Solvents of high dielectric constant minimise electrostatic attraction between cation and anion and promote dissociation[67].

Ionization and dissociation take place every time a macrozwitterion is formed. Therefore, both the solvating power and dielectric constant of the solvent will influence the polymerization rate. Strong solvating power will favour betaine formation, high dielectric constant chain growth.

As no author has systematically investigated the role of solvent, at this time it is not possible to give clear cut examples of the influence of these two solvent properties on a macrozwitterion polymerization. However, two observations are worth noting. Ogawa and Romero[16] have shown that acrylonitrile adds to the acrylonitrile-triethyl phosphite betaine at a rate approaching the propagation rate when the solvent is dimethylformamide, but more slowly in acetonitrile. The dielectric constants of these two solvents are virtually identical, but when compared on the Gutmann donor scale[67], dimethylformamide is seen to have a considerably higher solvating power for cations. It would appear that in acetonitrile the betaine is not fully ionized. Acetonitrile solvates anions more effectively than DMF. However, while this solvation shell will promote ionization it will also hinder the approach of monomer.

Johnston and Pepper[35] have found similar behaviour in another anionic polymerization. When dichloromethane was substituted for tetrahydrofuran the rate of amine initiated cyanoacrylate polymerizations fell markedly. The dielectric constants of these two solvents are similar. However, THF is a good donor, weak charge acceptor solvent, methylene chloride the reverse.

Nitromethane and nitrobenzene have an even more pronounced effect on the rate, increasing it by at least two orders of magnitude. Both are very poor donors, but do have high dielectric constants. Apparently, there is a dissociative step when cyanoacrylate is polymerized by aliphatic amines in THF.

Both Saegusa[63] and Schmidt[52] have shown that macrocycles are more likely to be formed when charge cancellation polymerizations are carried out in solvents of low dielectric constant. The equilibrium concentration of cyclic ion pairs will increase at the expense of linear ion pairs and free ions if the dielectric constant of the medium falls. Naturally under these circumstances there is a relative increase in the rate of macrocycle formation.

3.5 Zwitterion Propagation

The nature of ion pairing in macrozwitterion solutions has been a subject of speculation for several authors. Two types of ion pair have been proposed, mono- or bimolecular.

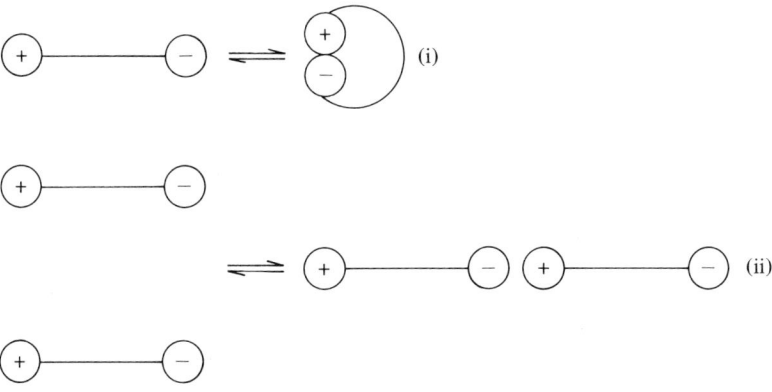

During vinyl polymerizations, the zwitterion concentration is usually very small. Ion pairing of type (ii) would be entropically very unfavourable. However, the formation of macrocycles during charge cancellation polymerizations suggests that at least in the case of low molecular weight zwitterions, cyclic ion pairs have more than just a fleeting existence.

A different situation exists with bipolar monomers. At the start of polymerization the active centre and monomer concentrations are equal. Both types of ion pairing are well documented.

4 Conclusions

In the light of past developments it seems to be appropriate at this point to try and identify areas which could reward further research in this field.

4.1 Anionic Polymerization

The ideal addition polymerization is one in which there are no termination or transfer reactions, initiation (betaine formation) is instantaneous and k_p lies between 10^2 and 10^4 $l \cdot mol^{-1} \cdot s^{-1}$. Such polymerizations can be conveniently followed by adiabatic calorimetry. At very high initiator concentrations it should be possible to measure the rates of the first monomer additions and isolate zwitterion oligomers for study. Cyanoacrylate polymerizations initiated by phosphine satisfy all these conditions except the last. k_p is 10^5 $l \cdot mol^{-1} \cdot s^{-1}$ at $-78\,°C$. However, cyanoacrylates with more bulky alkyl substituents may have smaller k_p's. It is possible that methylene malonate polymerizations may fulfil the above requirements.

Macrozwitterion polymerizations can be initiated by polymers with basic nitrogen or phosphorus atoms, provided these atoms are not part of the chain backbone[68]. Cyanoacrylates have been polymerized by poly(vinylpyridine)[34] and a styrene, diphenyl-4-vinylphenyl phosphine copolymer[69]. Polyampholite graft copolymers are formed. If the polymerization of a monomer-polymeric initiator combination also satisfies the criteria of an ideal macrozwitterion polymerization, each site on the polymer will be grafted and the graft length will depend solely on the monomer-initiator ratio. The anionic graft ends are sites for further reaction. Johnston and Pepper[34] converted cyanoacrylate anions to carboxylate with propiolactone and found that the product strongly co-ordinates metal atoms. Termination by acid yielded a polyelectrolyte. Clearly, polymerizations of this type have a considerable potential as a means for synthesizing graft copolymers.

4.2 Cationic Polymerization

Unequivocal demonstration of the formation of macrozwitterions is confined to anionic polymerization. Ironically zwitterions were first postulated as intermediates in cationic vinyl polymerizations initiated by Friedel-Crafts halides[4-6]. Friedel-Crafts halides, probably the most widely used cationic initiators, are molecules, not ions. However, formation of an anion with a metal-carbon bond seems to be energetically unfavourable and initiation is thought to occur by self ionization or involve a co-catalyst.

Only in two reports are there claims that the chains formed in a cationic polymerization are exclusively macrozwitterions: when formaldehyde is polymerized by BF_3[51] and vinyl ethers by $TCNQ$[31]. Neither of these polymerizations fit the criteria of an ideal macrozwitterion polymerization set out above. If cationic polymerizations of this type are to be successfully studied an initiator system must be developed which will generate macrozwitterions in a simple fashion from nucleophilic monomers. The organic chemical literature indicates the form such an initiator could take.

Gresham and Westfahl[70] found that benzene could be alkylated with methylenemalononitrile complexed by aluminium trichloride.

$$C_6H_6 + H_2C=C(CN \cdot AlCl_3)_2 \longrightarrow C_6H_5-CH_2-CH(CN)_2$$

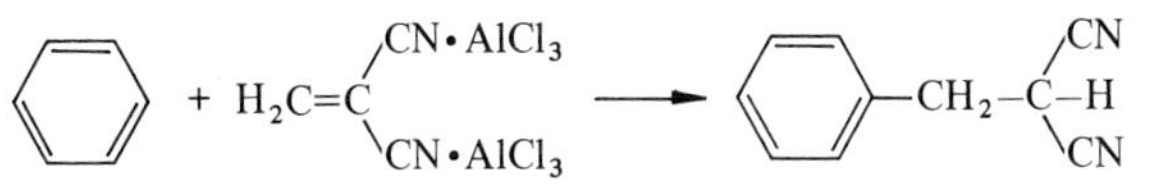

The first step in this reaction is probably betaine formation.

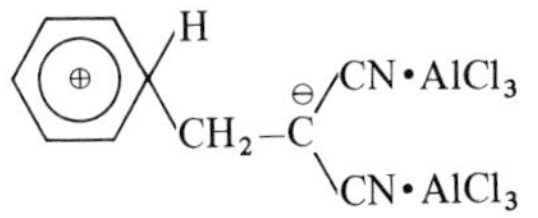

$$[C_6H_6-H]^{\oplus}-CH_2-\overset{\ominus}{C}(CN \cdot AlCl_3)_2$$

The polymerization of tetrahydrofuran in the presence of methylenemalononitrile has already been described. Co-ordination of aluminium trichloride to the cyano groups would make methylenemalononitrile an even more effective initiator of cationic polymerization.

With this in mind Johnston[68] synthesized the alkene shown below and studied its interaction with aluminium trichloride.

$$\text{4-}CH_3O\text{-}C_6H_4\text{-}CH=C(CN)_2 + 2AlCl_3 \longrightarrow \text{4-}CH_3O\text{-}C_6H_4\text{-}\overset{\oplus}{C}H\text{-}\overset{\ominus}{C}(CN \cdot AlCl_3)_2 \quad ?$$

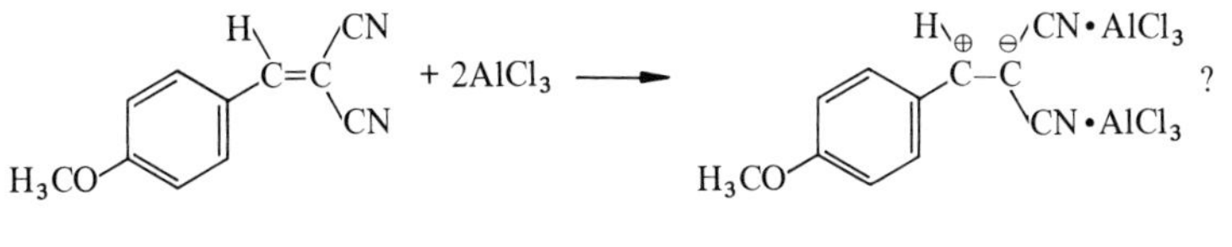

When methylene chloride solutions of the alkene and aluminium trichloride were mixed a yellow solid precipitated. The precipitate contained both alkene and aluminium trichloride. Except for the cyano absorption the IR spectrum of the alkene in the precipitate is unchanged. Aluminium trichloride has co-ordinated to the cyano groups but not broken the double bond – a betaine has not been formed. The complex appears to be polymeric, it will not dissolve in polar aprotic solvents. However, it will dissolve in chloroform without chemical reaction if a little methanol is added. Apparently, the polymeric structure is disrupted by methanol co-ordination. The ^1H NMR (in $CDCl_3$) spectrum shows that one molecule of alkene dissolves per molecule of alcohol. Despite the proximity of an alcohol molecule to the strongly polarized alkene no chemical reaction takes place in this solvent.

Clearly there is a large energy barrier to be overcome if a betaine with a carbonium ion is to be formed. The complex reacts with pure alcohol and the NMR spectrum of the products shows that the double bond has been broken.

Dicarbonyl ligands are less prone to form polymeric complexes, and if a complex between aluminium trichloride and the ligand, shown below, was soluble it would be easier to assess as an initiator.

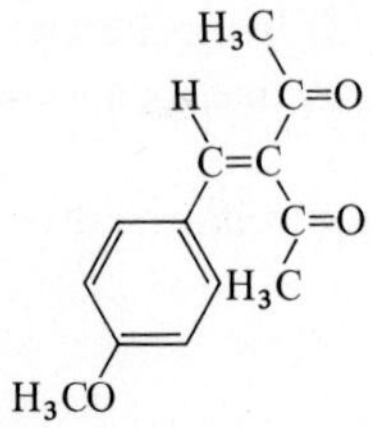

However, if initiation is to be rapid a strongly polarized monomer and polar solvent would seem to be necessary.

4.3 Charge Cancellation Polymerization

Frequently, macrocycles form a substantial proportion of the product of charge cancellation polymerizations. De Sarlo et al.[71] have shown that if such products have accessible donor atoms they can function as macrocyclic ligands. Other macrocycle forming monomers with donor atoms or groups would be of value.

Acknowledgements. The author would like to thank the following individuals who either helped in the gathering together of the publications cited or translated articles in French or German. Prof. *D. C. Pepper*, Drs. *D. J. Dunn, A. Frolich, J. M. Kelly*, and *J. M. Rooney*.

During the compilation and writing of this review the author was supported by funds from the *Irish National Board for Science and Technology*, Loctite (Ireland) Ltd. and the *Wellcome Trust*.

5 References

1. Ayres, D. C.: "Carbanions in Synthesis", London: Oldbourne 1966, p. 112
2. Walker, B. J.: "Organophosphorus Chemistry", Penguin 1972, p. 67
3. Gresham, T. L., Jansen, J. E., Shaver, F. W., Bankert, R. A., Fiedorek, F. T.: J. Am. Chem. Soc. 73, 3168 (1951)
4. Hunter, W. H., Yohe, R. V.: J. Am. Chem. Soc. 55, 1248 (1933)
5. Williams, H.: J. Chem. Soc. 1940, 775
6. Price, C. C.: "Reactions at Carbon-Carbon Double Bonds", Interscience, New York, 1946
7. Horner, L., Jurgeleit, H., Klupfel, K.: Justus Liebigs Ann. Chem. 59, 108 (1955)
8. Swarc, M.: Makromol. Chem. 35, 132 (1960)
9. Klippert, H., Ringsdorf, H.: Preprints I.U.P.A.C. Helsinki, Symposium on Macromolecules, 1–35, 211 (1972)
10. Klippert, H., Ringsdorf, H.: Makromol. Chem. 153, 289 (1972)
11. Ranogayets, F., Kotchetov, Ye. V., Markevich, M. A., Enikolopyan, N. S.: Pol. Sci. U.S.S.R. 15, 1508 (1973)
12. Ranogajec, F., Kotchetov, E. V. Markevich, M. A., Enikolopyan, N. S.: J. Polym. Sci., Polym. Symp. 42, 531 (1973)

13. Markevich, M. A., Kotchetov, E. V., Ranogajec, F., Enikolopyan, N. S.: J. Macromol. Sci. Chem. *8*(2), 265 (1972)
14. Ranogajec, F., Kotchetov, E. V., Markevich, M. A., Enikolopyan, N. S., Dvornik, I.: Croatica Chimica Acta *46*(2), 83 (1974)
15. Ogawa, T., Taninaka, T.: J. Polym. Sci. Part A-1, *10*, 2005 (1972)
16. Ogawa, T., Romero, J.: Eur. Polym. J. *13*, 419 (1977)
17. Ogawa, T., Quintana, P.: J. Polym. Sci. *13*, 2517 (1975)
18. Kotchetov, Ye. V., Berlin, A. A., Enikolopyan, N. S.: Pol. Sci. U.S.S.R. *8*, 1122 (1966)
19. Kotchetov, Ye. V., Berlin, A. A., Enikolopyan, N. S.: Pol. Sci. U.S.S.R. *8*, 1117 (1966)
20. Kotchetov, Ye. V., Berlin, A. A., Masal'skaya, Ye. M., Enikolopyan, N. S.: Pol. Sci. U.S.S.R. *12*, 1264 (1970)
21. Markevich, M. A., Kotchetov, Ye. V., Enikolopyan, N. S.: Pol. Sci. U.S.S.R. *13*, 1163 (1971)
22. Jaacks, V., Eisenbach, C. D., Kern, W.: Makromol. Chem. *161*, 139 (1972)
23. Saegusa, T., Kobayashi, S. Kimura, Y.: Macromolecules *7*, 257, (1974)
24. Vofsi, D., Katchalsky, A.: J. Polym. Sci. *26*, 127 (1957)
25. Grodzinsky, J., Katchalsky, A., Vofsi, D.: Makromol. Chem. *44–46*, 591 (1961)
26. Jagur-Grodzinsky, J.: "Encyclopedia of Polymer Science and Technology", Vol. 9, p. 315
27. Grodzinsky, J., Thesis, Ph. D.: Hebrew University 1961
28. Butler, G. B., Sivaramakrishnan, K. N.: Polym. Prepr. *17*(2), 608 (1976)
29. Hopf, H., Lüssi, H., Allisson, S.: Makromol. Chem. *44–46*, 95 (1961)
30. Jaacks, V., Franzmann, G.: Makromol. Chem. *143*, 283 (1971)
31. Donnelly, E. F., Johnston, D. S., Pepper, D. C., Dunn, D. J.: J. Polym. Sci., Polym. Lett. Ed. *15*, 399 (1977)
32. Pepper, D. C.: J. Polym. Sci., Polym. Symp. *62*, 65 (1978)
33. Johnston, D. S., Pepper, D. C.: Makromol. Chem. *182*, 393 (1981)
34. Johnston, D. S., Pepper, D. C.: Makromol. Chem. *182*, 407 (1981)
35. Johnston, D. S., Pepper, D. C.: Makromol. Chem. *182*, 421 (1981)
36. Oguni, N., Kamachi, M., Stille, J. K.: Macromolecules *7*, 435 (1974)
37. Tarvin, R. F., Aoki, S., Stille, J. K.: Macromolecules *5*, 663, (1972)
38. Machacek, Z., Mejzlik, J., Pac, J.: J. Polym. Sci. *52*, 309 (1961)
39. Enikolopyan, N. S.: J. Polym. Sci. *58*, 1301 (1962)
40. Mathes, N., Jaacks, V.: Makromol. Chem. *135*, 49 (1970)
41. Künzel, E., Giefer, A., Kern, W.: Macromol. Chem. *96*, 17 (1966)
42. Etienne, Y., Soulas, R.: J. Polym. Sci., Part C, *4*, 1061 (1963)
43. Jaacks, V., Mathes, N.: Makromol. Chem. *131*, 295 (1970)
44. Mathes, N., Jaacks, V.: Makromol. Chem. *142*, 209 (1971)
45. Mayne, N. R.: A.C.S. Advan. Chem. Ser. *29*, 175 (1973)
46. Wilson, D. R., Beaman, R. G.: J. Polym. Sci., Part A-1, *8*, 2161 (1970)
47. Fisher, R. F.: J. Polym. Sci. *44*, 155 (1960)
48. Wieland, Th.: Angew. Chem. *66*, 507 (1954)
49. Kricheldorf, H. R., Bosinger, K.: Makromol. Chem. *177*, 1243 (1976)
50. Seeney, C. E., Harwood, H. J.: "Ionic Polymerisation, Unsolved Problems", Furukawa, J. and Vogl, O. Eds., Dekker, New York 1976, p. 144
51. Kern, W., Jaacks, V.: J. Polym. Sci. *48*, 399 (1960)
52. Schmidt, D. L., Smith, H. B., Yoshimine, M., Hatch, M. J.: J. Polym. Sci., Part A-1, *10*, 2951 (1972)
53. Chapiro, A., Du Lieu, J., Laborie, F.: Paper presented at 3rd International Symposium on Radiation Chemistry, Tihany 1971
54. Schmidt, D. L.: Polym. Prepr. *18*(1), 121 (1977)
55. Brandi, A., De Sarlo, F., Guarna, A.: J. Chem. Soc., Perkin I *1827* (1976)
56. Kantor, S. W., Osthoff, R. C.: J. Am. Chem. Soc. *75*, 931 (1953)
57. Hamann, K.: Z. Elektrochem. *60*, 317 (1956)
58. Huisgen, R.: Angew. Chem. *67*, 439 (1955)
59. Bawn, C. E. H., Ledwith, A., Matthies, P.: J. Polym. Sci. *34*, 93 (1959)
60. Plesch, P. H.: Ricerca Sci. *25*, 140 (1955)
61. Saegusa, T.: Chem. Technol. *5*, 295 (1975)
62. Saegusa, T., Kobayashi, S., Yokoyama, T.: Polym. Prepr. *18*(1), 125 (1977)

63. Saegusa, T., Kobayashi, S., Hayashi, K.: Macromolecules *11*, 360 (1978)
64. Overberger, C. G.: "Encyclopedia of Polymer Science and Technology", Vol. 11, p. 96
65. Swarc, M.: "Carbanions, Living Polymers and Electron Transfer Processes", Interscience New York 1968, p. 596
66. Walker, B. J.: "Organophosphorus Chemistry", Penguin 1972, p. 49
67. Mayer, U.: Pure Appl. Chem. *41*, 291 (1975)
68. Johnston, D. S., Thesis, Ph. D., University of Dublin 1978
69. Miura, M., Akatsu, F., Ito, H., Nagakuko, K.: J. Polym. Sci., Polym. Chem. Ed. *17*, 1568 (1979)
70. Westfahl, J. C., Gresham, T. L.: J. Am. Chem. Soc. *76*, 1076 (1954)
71. De Sarlo, F., Guarna, A., Speroni, A.: Chem. Abstr. *88*, 89639c (1978)

Received December 19, 1980
W. Kern (Editor)

Ring-Opening Polymerization
of Atom-Bridged and Bond-Bridged Bicyclic Ethers, Acetals and Orthoesters

Y. Yokoyama and H. K. Hall, Jr.

Department of Chemistry, College of Liberal Arts, University of Arizona, Tucson, AZ 85721, USA

During these two decades, the polymerization of bicyclic compounds has been studied to obtain basic knowledge on ring-opening polymerization and for industrial and biomedical applications. In 1961, Korshak et al.[1] reported the stereospecific polymerization of 1,6-anhydro-2,3,4-tri-O-methyl-D-glucopyranose. Schuerch and coworkers[2-3] have developed the chemical synthesis of polysaccharides of biomedical interest. Very recently, Sumitomo and Okada[4] reviewed the reaction mechanism of the ring-opening polymerization and the structures and properties of the resulting polymers of bicyclic acetals, bicyclic oxalactones, bicyclic oxalactams, and related heterobicyclo compounds. Tadokoro[5, 6] reported the structure of crystalline polymers by X-ray diffraction, infrared, Raman spectroscopy, and energy calculations. Many workers[7] have studied the structure of oxygen-containing polymers in solution. Finally, the anomeric effect and gauche effect have been made clear by many workers[8, 9].

Therefore, we hope to discuss the polymerization mechanism and polymerizability of bicyclic compounds containing oxygen atoms and their relation to the monomer and polymer structures. Finally, some biomedical application of polymers will be mentioned.

The review will be limited to atom- and bond-bridged bicyclic monomers. The important work of Bailey on the polymerization of spiro bicyclo orthoesters and spiro bicyclic orthocarbonates has been adequately described elsewhere[10a-d, 60-63]. The outstanding and systematic body of work by Schuerch[2, 3] and others on the ring opening polymerization of bicyclic acetals derived from carbohydrate precursors will also not be covered here.

1 **Polymerization of Bicyclic Compounds Having Oxygen Atoms** 109
  Polymerizabilities of bicyclic compounds are listed in Table 1. From these results, the polymerizability of unsaturated bicyclic compounds is summarized in Table 2.

2 **Thermodynamics** 115
  Thermodynamical data are shown. Thermodynamical polymerizability is discussed using this data along with the strain energies.

3 **Monomer Conformation** 119
  Conformational analyses of monomers are shown. Gauche and anomeric effects are discussed.

4 **Polymer Conformation** 123
  Conformational analyses of polymers are shown. Gauche and anomeric effects are also discussed.

5 **Kinetics** 127
Kinetic data are shown. From the data, monomer reactivities are compared to monocyclic compounds.

6 **Polymerization Mechanism and Stereoselectivity** 129
Polymerization mechanism and stereoregular propagation are discussed and summarized in Table 6.

7 **Medical Application of Polymers** 133
The possibility of biomedical application of polymers is mentioned.

8 **References** 134

1 Polymerization of Bicyclic Compounds Having Oxygen Atoms

Many kinds of bicyclic ethers, acetals, and orthoesters have been cationically polymerized. Almost all bicyclic compounds have shown polymerizability. A few monomers give dimer or higher oligomers instead of polymers. The results are listed in Table 1. The polymerizability for unsubstituted bicyclic compounds is summarized in Table 2.

Table 1. Polymerization of bicyclic ethers, acetals, and orthoesters

	Monomer	Polymerizability	Property of polymer	Ref.
Bicyclic ethers				
[2.2.1] Series				
7-Oxabicyclo[2.2.1]heptane		+	1,4-trans configuration for cyclohexane ring (white powder, m.p. 450°, $[\eta]_{inh}$ = 1.04 dl/g[13])	11–14, 16, 17, 26
exo-2-Methyl-7-oxabicyclo[2.2.1]heptane	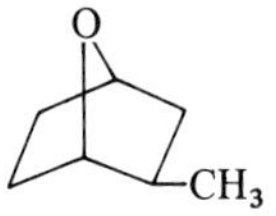	+	1,4-trans configuration for cyclohexane ring[13, 20] (m.p. 200°[13], $[\eta]_{inh}$ = 0.85 dl/g[27])	13, 15, 16, 18–22, 24, 26, 27
endo-2-Methyl-7-oxabicyclo[2.2.1]heptane	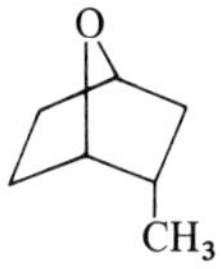	+	1,4-trans configuration for cyclohexane ring[13, 20] (m.p. 257°[13], $[\eta]_{inh}$ = 1.97 dl/g[27])	13, 15, 16, 20, 22–24, 26, 27
exo-2-tert-Butyl-7-oxabicyclo[2.2.1]heptane	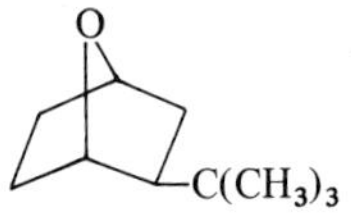	+	cis, trans(a,e,a) configuration for cyclohexane ring $\overline{M_n}$ = 480[24]	24, 25
endo-2-tert-Butyl-7-oxabicyclo[2.2.1]heptane	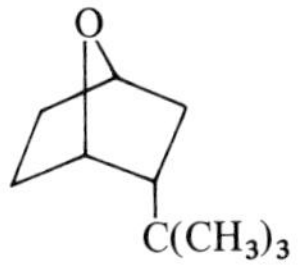	+	trans, trans(e,e,e) configuration for cyclohexane ring $\overline{M_n}$ = 660[24]	24, 25
endo,exo-2,6-Dimethyl-7-oxabicyclo[2.2.1]heptane	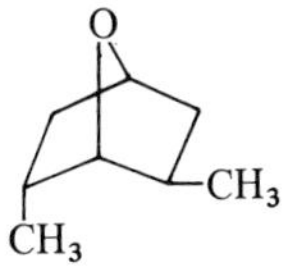	+	highly crystalline polymer $[\eta]_{inh}$ = 0.54 dl/g	27
exo,exo-2,6-Dimethyl-7-oxabicyclo[2.2.1]heptane	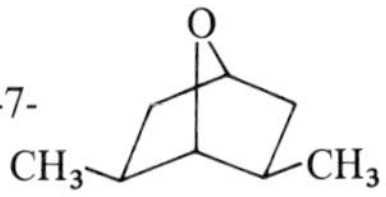	–		27
exo,exo-2,3-Dimethyl-7-oxabicyclo[2.2.1]heptane	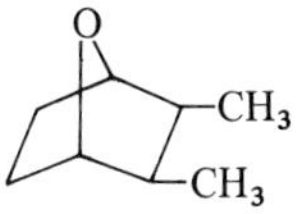	+	white powder	28

Table 1. (continued)

Monomer		Polymerizability	Property of polymer	Ref.
endo,exo-2,3-Dimethyl-7-oxabicyclo[2.2.1]heptane	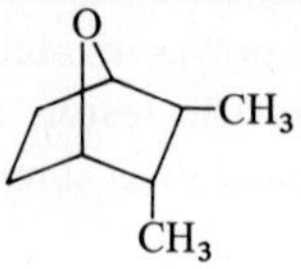	+	trans, trans, trans configuration for cyclohexane ring, white powder	28)
endo,endo-2,3-Dimethyl-7-oxabicyclo[2.2.1]heptane	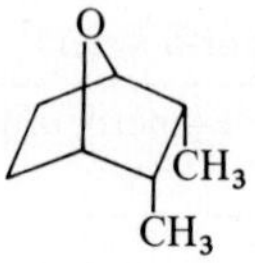	+		28)
exo,exo-2,5-Dimethyl-7-oxabicyclo[2.2.1]heptane	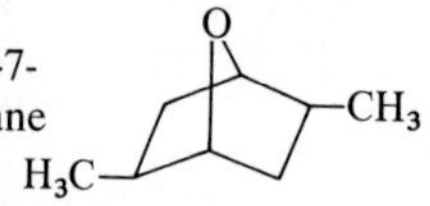	+	waxy polymer	28)
endo,exo-2,5-Dimethyl-7-oxabicyclo[2.2.1]heptane	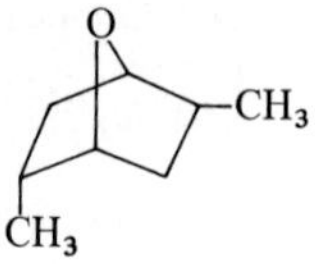	+	highly crystalline polymer	28)
endo,endo-2,5-Dimethyl-7-oxabicyclo[2.2.1]heptane	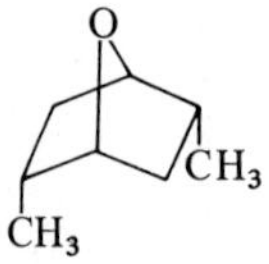	+	waxy polymer	28)
2,3-Benzo-7-oxabicyclo[2.2.1]heptane	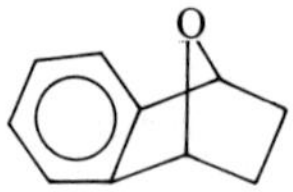	+	white powder, melted at 50° to a viscous liquid	13)
1,4-Endoxadecalin	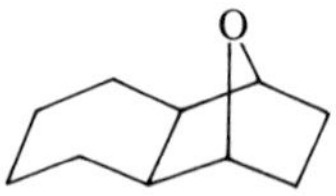	–		13)
1,4-Endoxa-5,8-methanodecalin	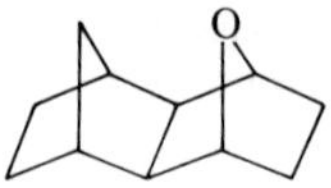	–		13)

[3.2.1] Series, etc.

Monomer		Polymerizability	Property of polymer	Ref.
6-Oxabicyclo[3.2.1]octane	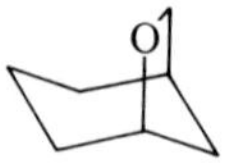	+[a]		13)
3-Oxabicyclo[3.3.1]nonane	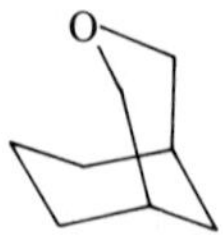	–		13)
2-Oxabicyclo[2.2.2]octane	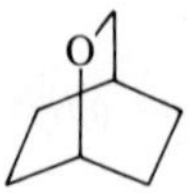	+	1,4-trans configuration for cyclohexane ring[29], a sticky solid	13, 29)

Table 1. (continued)

Monomer		Polymerizability	Property of polymer	Ref.
3-Oxabicyclo[3.2.2]nonane	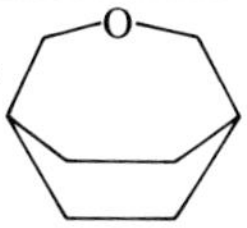	–		35)
trans-2-Oxabicyclo[3.3.0]octane	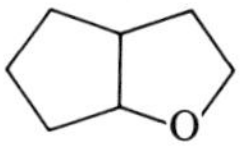	+	kinetic study	30)
trans-3-Oxabicyclo[3.3.0]octane	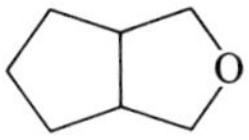	+	kinetic study	30)
cis-3-Oxabicyclo[3.3.0]octane	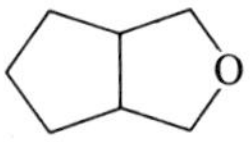	–		30)
trans-7-Oxabicyclo[4.3.0]nonane	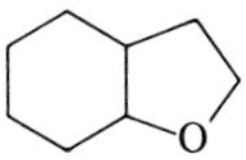	+	1,2-trans configuration for cyclohexane ring; elastomeric solid polymer $T_g \sim 0°$, $\overline{M_n} = 53\,700$ $\overline{M_w}/\overline{M_n} = 1.78$	31)
8-Oxabicyclo[4.3.0]nonane	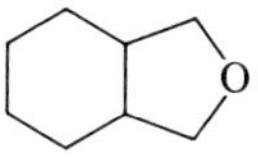	+	a thick syrupy polymer $\overline{M_n} = 5700 \sim 9800$	13)
trans-8-Oxabicyclo[4.3.0]nonane	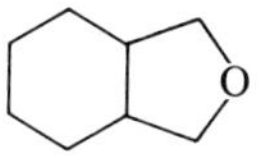	+	$\overline{M_n} = 200\,000$	32)
cis-8-Oxabicyclo[4.3.0]nonane	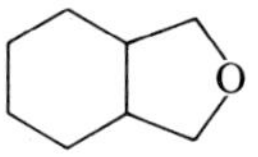	±	no homopolymerizability copolymerized with trans-monomer	32)
7-Oxabicyclo[4.1.0]heptane	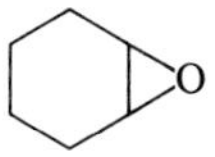	+	powdery linear amorphous substance, $T_g = 59°$ [36)	33, 36–39)
7-Oxabicyclo[4.2.0]octane	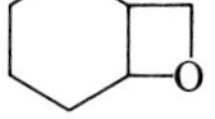	+	m.p. 165°	34)
	Bicyclic acetals			
2,6-Dioxabicyclo[2.2.1]heptane	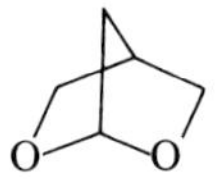	+	60:40% mixture of tetrahydrofuran isomers (probably trans:cis) $[\eta]_{inh} = 0.35$ dl/g liquid (rubber)	40)
2,7-Dioxabicyclo[2.2.1]heptane	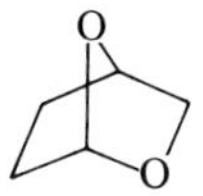	+	tetrahydrofuran structure $[\eta]_{inh} = 0.09–0.45$ dl/g	40)

Table 1. (continued)

Monomer		Polymerizability	Property of polymer	Ref.
1,3,3-Trimethyl-2,7-dioxabicyclo[2.2.1]heptane	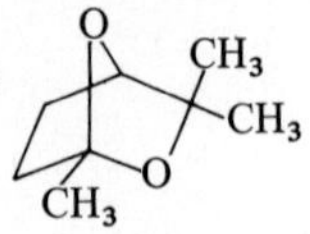	+	tetrahydropyran (more than 90%) and tetrahydrofuran mixture. $\overline{Mn} = 400 \sim 600$	41)
6,8-Dioxabicyclo[3.2.1]octane	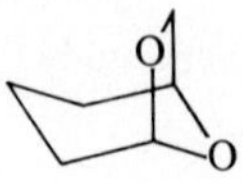	+	trans configuration for tetrahydropyran ring ($[\eta]_{inh} = 1.87$ dl/g, melted at 160–180°)	42–49)
1-Methyl-6,8-dioxabicyclo[3.2.1]octane	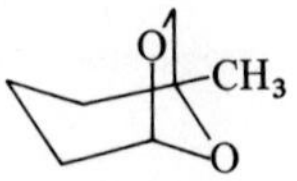	+	m.p., 94–97°	42)
5,7-Dimethyl-6,8-dioxabicyclo[3.2.1]octane	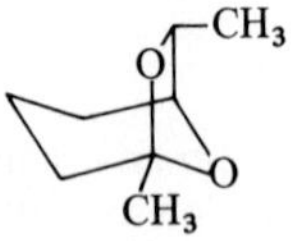	+	m.p., 102–105°	42)
4-Bromo-6,8-dioxabicyclo[3.2.1]octane	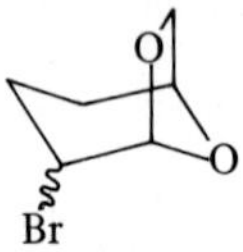	+	$[\eta]_{inh} = 0.10$ dl/g[4]	4, 45)
3,6,8-Trioxabicyclo[3.2.1]octane	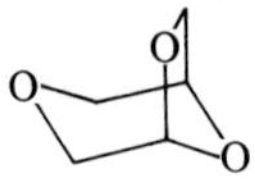	+	cis and trans configuration for tetrahydropyran ring $[\eta]_{inh} = 0.56 \sim 0.80$ dl/g	45)
6,8-Dioxabicyclo[3.2.1]oct-3-ene	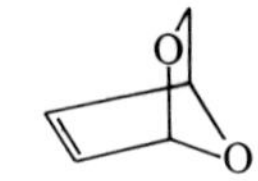	+	dihydropyran structure $\overline{Mn} = 6400$	50–53)
2,6-Dioxabicyclo[2.2.2]octane	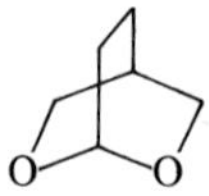	+	trans configuration for tetrahydropyran ring $[\eta]_{inh} = 0.13–1.1$ dl/g	54)
cis-7,9-Dioxabicyclo[4.3.0]nonane	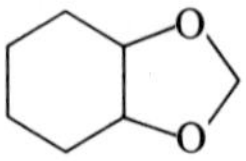	±	cyclic dimer	55)
trans-7,9-Dioxabicyclo[4.3.0]nonane	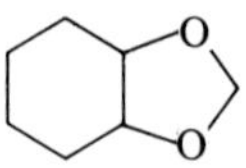	+	cyclic dimer and polymer, $\overline{Mn} = 5300 \sim 10\,100$ (GPC values)	55)

Bicyclic orthoesters

Monomer		Polymerizability	Property of polymer	Ref.
2,6,7-Trioxabicyclo[2.2.1]heptane	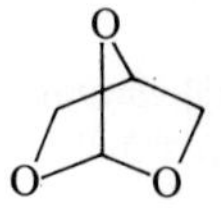	+	cis and trans configuration for 1,3-dioxolane ring at low temperature, two ring-opened structure at 80 °C	56, 57)

Table 1. (continued)

Monomer		Polymerizability	Property of polymer	Ref.
2,7,8-Trioxabicyclo[3.2.1]octane	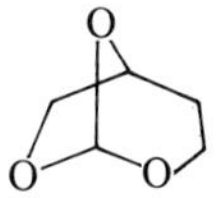	+	cis and trans configuration for 5-, 6- and 7-membered units	58)
2,8,9-Trioxabicyclo[3.3.1]nonane	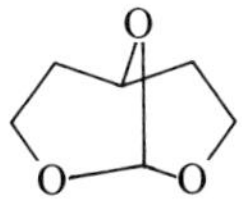	±	cyclic dimer and oligomers	58)
2,6,7-Trioxabicyclo[2.2.2]octane	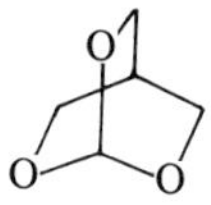	+	highly crystalline polymer	59)
4-Methyl-2,6,7-trioxabicyclo[2.2.2]octane	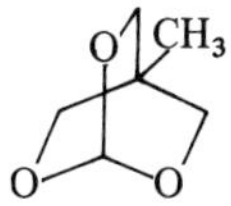 CH$_3$	+	highly crystalline polymer	59)
4-Ethyl-2,6,7-trioxabicyclo[2.2.2]octane	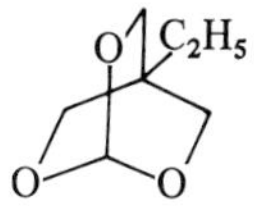 C$_2$H$_5$	+	two ring-opened structure (polyether with ester substituents)	60, 61)
4-Bromomethyl-2,6,7-trioxabicyclo[2.2.2]octane	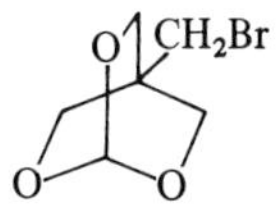 CH$_2$Br	+	highly crystalline polymer	59)
4-Hydroxymethyl-2,6,7-trioxabicyclo[2.2.2]octane	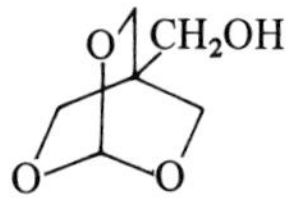 CH$_2$OH	+	highly crystalline polymer	59)
4-Nitro-2,6,7-trioxabicyclo[2.2.2]octane	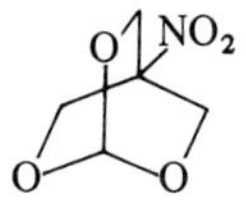 NO$_2$	+	highly crystalline polymer	59)
4-Amino-2,6,7-trioxabicyclo[2.2.2]octane	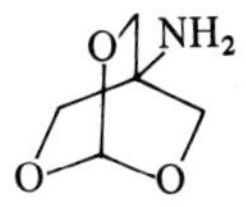 NH$_2$	—		59)
4-[N,N-dimethylamino]-2,6,7-trioxabicyclo[2.2.2]octane	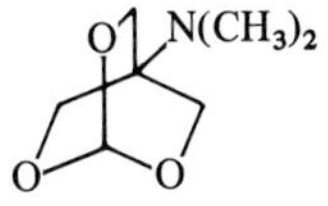 N(CH$_3$)$_2$	—		59)
4-Acetylamino-2,6,7-trioxabicyclo[2.2.2]octane	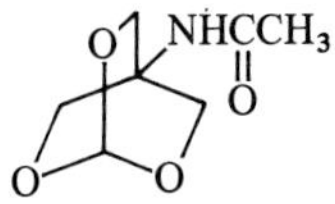 NHCCH$_3$ (O)	+	poly-orthoester powder	59)
4-Carboethoxy-2,6,7-trioxabicyclo[2.2.2]octane	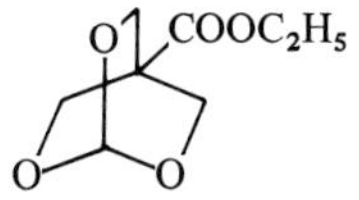 COOC$_2$H$_5$	+	highly crystalline poly-orthoester	59)

Table 1. (continued)

Monomer		Polymerizability	Property of polymer	Ref.
4-p-Toluenesulfonyl-oxymethyl-2,6,7-trioxabicyclic[2.2.2]octane	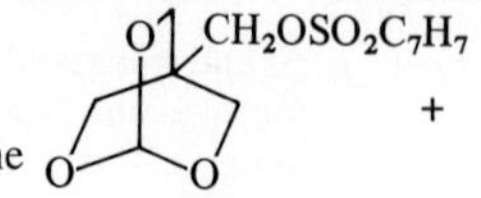 $CH_2OSO_2C_7H_7$ structure	+	highly crystalline poly-orthoester	59)
1,4-Dimethyl-2,6,7-trioxabicyclo[2.2.2]octane	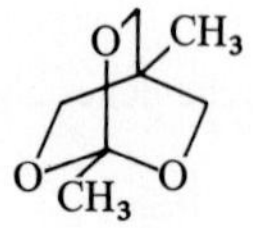 CH_3 / CH_3 structure	+	highly crystalline poly-orthoester	59)
1,4-Diethyl-2,6,7-trioxabicyclo[2.2.2]octane	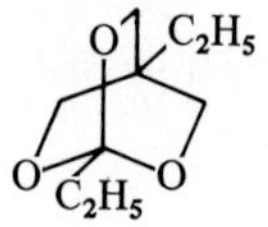 C_2H_5 / C_2H_5 structure	+	two ring-opened structure (polyether with ester substituent)	61, 62)
1-Ethyl-4-hydroxymethyl-2,6,7-trioxabicyclo[2.2.2]octane	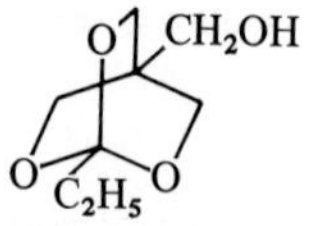 CH_2OH / C_2H_5 structure	+	two ring-opened structure (polyether with ester substituent)	62)
4-Ethyl-1-phenyl-2,6,7-trioxabicyclo[2.2.2]octane	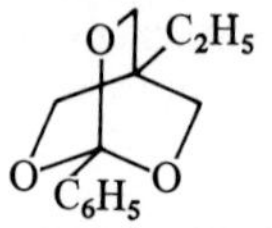 C_2H_5 / C_6H_5 structure	+	two ring-opened structure (polyether with ester substituent)	61, 62)
1-Vinyl-4-ethyl-2,6,7-trioxabicyclo[2.2.2]octane	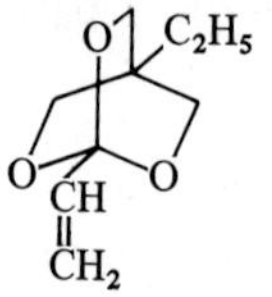 C_2H_5 / CH=CH_2 structure	+	radical polymerization	63)
4-Hydroxymethyl-1-methyl-2,6,7-trioxabicyclo[2.2.2]octane	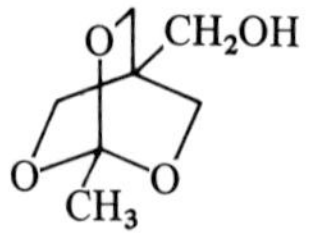 CH_2OH / CH_3 structure	+	highly crystalline poly-orthoester	59)
4-Carbomethoxy-1-methyl-2,6,7-trioxabicyclo[2.2.2]octane	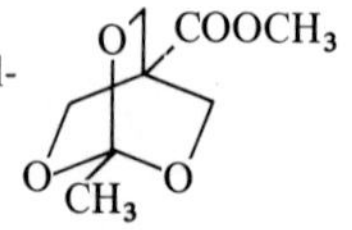 $COOCH_3$ / CH_3 structure	+	highly crystalline poly-orthoester	59)
4-Carboethoxy-1-methyl-2,6,7-trioxabicyclo[2.2.2]octane	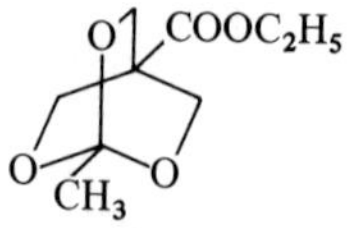 $COOC_2H_5$ / CH_3 structure	+	highly crystalline poly-orthoester	59)
4-Nitro-1-methyl-2,6,7-trioxabicyclo[2.2.2]octane	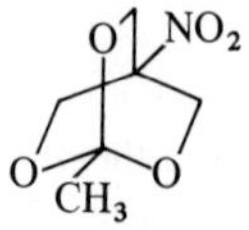 NO_2 / CH_3 structure	+	highly crystalline poly-orthoester	59)
4-Amino-1-methyl-2,6,7-trioxabicyclo[2.2.2]octane	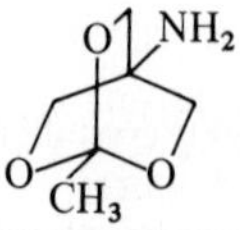 NH_2 / CH_3 structure	−		59)

Table 2. Polymerizability of unsubstituted bicyclic compounds

Monomer	Ring systems								
	[2.2.1]	[3.2.1]	[3.3.1]	[2.2.2]	[3.2.2]	[3.3.0]	[4.3.0]	[4.2.0]	[4.1.0]
Bicyclic ethers	+	+	+	+	−	+[a]	+[b]	+	+
Bicyclic acetals	+	+		+			+[c]		
Bicyclic orthoesters	+	+	±[d]	+					

[a] Ref. 30; cis-3-Oxabicyclo [3.3.0] octane does not polymerize
[b] Ref. 32; cis-8-Oxabicyclo [4.3.0] nonane has no homopolymerizability, but copolymerizes with trans monomer
[c] Ref. 55; cis-7,9-Dioxabicyclo [4.3.0] nonane gives cyclic dimer
[d] Ref. 58; Cyclic dimer and higher oligomers are formed

2 Thermodynamics

So far only a few thermodynamical studies for bicyclic monomers have been reported. They are listed in Table 3 along with the results for the polymerizations of monocyclic compounds for comparison. Plesch et al.[23, 26] reported the thermodynamical study for 7-oxabicyclo [2.2.1] heptane, endo- and exo-2-methyl-7-oxabicyclo [2.2.2] hexane using calorimetric measurements. The enthalpies of combustion of these polymers have been measured. From the literature values for the enthalpies of combustion of the liquid monomers, the enthalpies of the polymerization have been derived for three polymerizations. The entropies of polymerization have been estimated, yielding the calculated ceiling temperature of 320, 240, and 200 °C, respectively, for the three bicyclic ethers (Table 3). The results show that when the 2-methyl groups in the polymer are in equatorial positions, they do not contribute to the ring strain. When comparing the data for exo- and endo-2-methyl-7-oxabicyclo [2.2.1] hexane, the exo-monomer is thermodynamically more favored for the polymerization. On the other hand kinetically the endo-monomer is more reactive than the exo-monomer[25]. In comparison with data for the monocyclic ether tetrahydrofuran, the atom-bridged structure increases the ring strain (more favorable for polymerization).

The polymerization diagram of exo- and endo-2-methyl-7-oxabicyclo [2.2.1] heptane is shown in Fig. 1, indicating that the activation energies for the decomposition will be ca. 114 and 103 kJ/mol for exo- and endo-monomer, respectively.

Sumitomo and Okada[4] reported the thermodynamical data for the polymerization of 6,8-dioxabicyclo [3.2.1] octane. The ΔH_{ss} value is less negative (less favorable) than that for 1,3-dioxolane[68], suggesting that the [3.2.1] bicyclic structure does not increase the ring strain as would be expected from molecular model inspection and polymerization experiments[70]. The ΔS_{ss} value is less negative (more favorable) than for 1,3-dioxolane. This is the opposite of what would be predicted from thermodynamical considerations of

Table 3. Thermodynamical data for polymerization of mono- and bicyclic compounds

Monomers	Solvent	Temp. °C	$-\Delta H_{xy}$ kJ/mol	$-\Delta S_{xy}$ J/k·mol	Tc. °C	xy	Method[a]	Ref.
Ethers								
7-Oxabicyclo [2.2.1] heptane	–	25	44.3 ± 1.9	75.3	320	lc	C	26
exo-2-Methyl-7-oxabicyclo [2.2.1] heptane	–	25	49.7 ± 3.1	96.2	240	lc	C	26
endo-2-Methyl-7-oxabicyclo [2.2.1] heptane	–	25	48.4 ± 3.1	96.2	200	lc	C	23
trans-7-Oxabicyclo [4.3.0] nonane	–	−35 ~20	14.0	55.7	–	lc	E	31
Tetrahydrofuran	–	25	38 ± 4.2	–	–	lc	C	66
Tetrahydrofuran	bulk	25 ~80	19.2	74.1	84	ls	E	67
Acetals								
6,8-Dioxabicyclo [3.2.1] octane	CH_2Cl_2	−20 ~30	17.5 ± 1.7	59.4 ± 5.9	–	ss	E	55
trans-7,9-Dioxabicyclo [4.3.0] nonane	Toluene-d_8	25 ~105	24.1 ± 0.8	61.6	118	ss	E	55
1,3-Dioxolane	CH_2Cl_2	−88 ~15	21.3 ± 0.8	79.8	1.5[b]	ss	E	68
1,3-Dioxolane	Benzene	0 ~65	15.1	58.5	–	ss	E	69

[a] C, Calorimetric measurement, and E, equilibrium polymerization
[b] For 1 mol/l solution

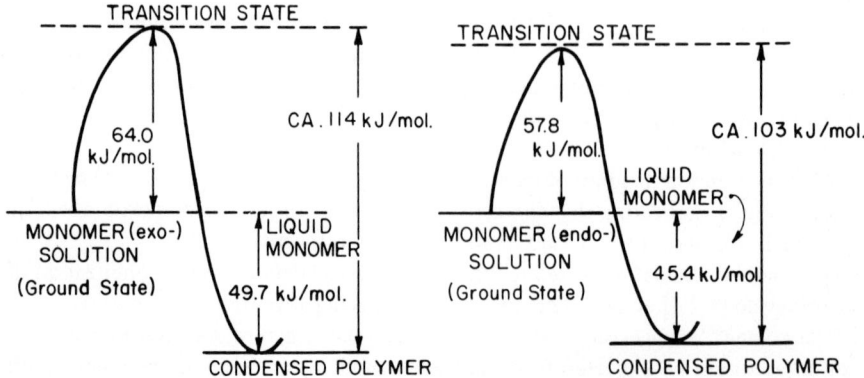

Fig. 1. Enthalpy change for the polymerization and depolymerization of exo- and endo-2-methyl-7-oxabicyclo[2.2.1]heptanes

cycloalkanes[71], if 6,8-dioxabicyclo [3.2.1] octane is regarded as a 2,4-dimethyl-substituted 1,3-dioxolane. However, if it is regarded as a 2,6-disubstituted oxepan, the difference between the equilibrium monomer concentrations at 10 °C for 6,8-dioxabicyclo [3.2.1] octane[5] and oxepane[72] (0.74 and 6×10^{-2} mol/l in dichloromethane respectively) corresponds to the thermodynamical consideration[71, 73].

Kops and Spanggaard[55] reported the thermodynamical study for trans-7,9-dioxabicyclo [4.3.0] nonane. Compared with the values obtained for the polymerization of 1,3-dioxolane in benzene[69], the ΔH_{ss} for trans-7,9-dioxabicyclo [4.3.0] nonane is larger, reflecting the greater strain in the dioxolane ring when it is fused in the trans-position to a six-membered ring. On the other hand the entropy change is comparable for the two compounds, which is to be expected, because the same sized ring is being opened in the polymerization[55]. If the monomer is regarded as a trans-4,5-dimethyl-1,3-dioxolane, the value of ΔH_{ss} and ΔS_{ss} seem to be opposed to the thermodynamical consideration[71, 73] and experimental results[74, 75], indicating that the ring strain (more negative ΔH_{ss}) due to the bicyclic structure contributes to the high thermodynamical reactivity of trans-7,9-dioxabicyclo [4.3.0] nonane.

Strain energies of some cyclic and bicyclic compounds are shown in Table 4. As reported by J. D. Cox and G. Pilcher[80], the strain energies of cyclopentane, tetrahydrofuran, and 1,3-dioxolane are very close, i.e. the replacement of $-CH_2-$ by O-atom in a cyclic compound has no significant effect on the conventional strain energy in these monocyclic compounds.

Table 4. Strain energies of some cyclic and bicyclic compounds

Compound	Strain energy, kJ/mol	Ref.
5–Ring		
Cyclopentane	25	76
Tetrahydrofuran	28	76
1,3-Dioxolane	31	76
6-Ring		
Cyclohexane	0.4	76
Tetrahydropyran	9.2	76
1,3-Dioxane	12	76
Bicyclic		
Bicyclo [2.2.1] heptane	71	77
Bicyclo [2.2.1] heptane	64	78
7-Oxabicyclo [2.2.1] heptane	58	79
exo-2-Methyl-7-oxabicyclo [2.2.1] heptane	60	79
endo-2-Methyl-7-oxabicyclo [2.2.1] heptane	64	79
Bicyclo [2.2.2] octane	54	77
Bicyclo [3.2.1] octane	51	77
Bicyclo [3.3.1] nonane	40	77
cis-Bicyclo [3.3.0] octane	52	77
8-Ring		
Cyclooctane	41	76

Tetrahydrofuran and 1,3-dioxolanes are conformationally labile like cyclopentane. The IR[81], dipole moment and microwave data[82] of tetrahydrofuran and IR of 1,3-dioxolane[81] show their pseudorotation has very small energy barriers between the envelope and twist conformers. MO studies of 1,3-dioxolane[83] show that a pseudorotation path is favored for conformational changes in 1,3-dioxolanes with the twist conformer slightly more stable than the envelope conformation. H-NMR study of 1,3-dioxolane in solution[84] also exhibits the twist conformation.

In tetrahydropyran[7, 8], the chair conformation is more stable than the boat conformation as well as in cyclohexane. The kinetic parameters of the ring inversion of tetrahydropyran were determined by Gratti, Segre, and Morandi[85], indicating that the conformational mobility of the ring is not seriously affected by the O-atom when compared to cyclohexane. 1,3-Dioxane is also similar to cyclohexane[86–88]. The X-ray measurements[89–91] and ^{1}H-NMR spectra[92–96] however, show that the ring is slightly puckered due to the difference of the bond lengths and the angles between the C–C–C and O–C–O. As a result the strain energy of 1,3-dioxane is higher than cyclohexane as shown in Table 4.

In contrast to the results for monocyclic compounds, in bicyclic compounds the rigidity of bridged structure frequently permits more substantial effects than in acyclic or monocyclic compounds. The bridging locks the cyclohexane ring into the boat and/or twist-bond conformation. For example, bicyclo [2.2.1] heptane and bicyclo [2.2.2] octane have one and two boat cyclohexane rings respectively. The enthalpies of the boat and twist-boat conformers of cyclohexane are 27 and 23 kJ/mol higher than that of the chair conformer. Therefore, bicyclo [2.2.1] heptane and bicyclo [2.2.2] octane have ring strains of 64 and 54 kJ/mol, respectively, as shown in Table 4.

J. D. Cox and G. Pilcher's statement[80] may be inapplicable to the discussion of the ring strain in bicyclic compounds having oxygen atoms. For example, the ring strain of 7-oxabicyclo [2.2.1] heptane is less than that of bicyclo [2.2.1] heptane as shown in Table 4. The difference of strain energies is explained in terms of the difference of the angle C_2-C_1-C_6, namely 113.5° for 7-oxabicyclo [2.2.1] heptane[97] and 108° for bicyclo [2.2.1] heptane[98]. That is, the nonbonded C_2-C_6 (and/or the corresponding endo-hydrogens) repulsion in 7-oxabicyclo [2.2.1] heptane is less than in bicyclo [2.2.1] heptane[77]. The ring strain of endo- and exo-2-methyl-7-oxabicyclo [2.2.1] heptanes is slightly higher than that of 7-oxabicyclo [2.2.1] heptane.

As mentioned before, the ΔH_{ss} value for the polymerization of 6,8-dioxabicyclo [3.2.1] octane is less negative than that for 1,3-dioxolane[5]. Bicyclo [3.2.1] octane, in which a five-membered ring is fused to a cyclohexane ring at two axial positions, has 51 kJ/mol strain energy. However, from the fact that 7-oxabicyclo [2.2.1] heptane has less strain energy than bicyclo [2.2.1] heptane can be deduced that the strain energy of 6,8-dioxabicyclo [3.2.1] octane may also be less than that of bicyclo [3.2.1] octane.

Thermodynamical studies of the polymerization of bicyclic orthoesters has not yet been done.

Symmetry Number (σ)

If the $-CH_2-$ is substituted by an O-atom in bicyclic compounds, the symmetry number changes. When the magnitudes of the enthalpy and entropy among a set of compounds are nearly comparable, the symmetry-number corrections are significant. The difference between II and IV is RT[ln(3) − ln(1)] = 9.2 TJ/mol. Hence, the entropy difference between −78 °C and 25 °C is 0.92 kJ/mol.

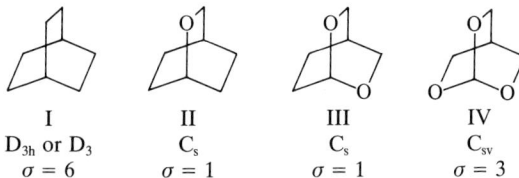

I	II	III	IV
D_{3h} or D_3	C_s	C_s	C_{sv}
$\sigma = 6$	$\sigma = 1$	$\sigma = 1$	$\sigma = 3$

For the set of bicyclo [2.2.1] monomers, V–VIII, the same entropy difference is only 0.58 kJ/mol. Therefore, at the normal temperature for the polymerization, the correction may be neglected.

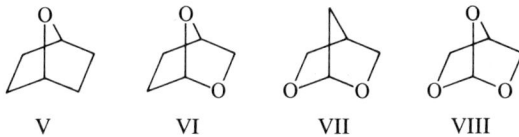

V   VI   VII   VIII

3 Monomer Conformation

Several comprehensive summaries of the conformation of monocyclic compounds[8, 77, 99, 100–101] and bicyclic compounds[77, 103] are available. However, only a few conformational analyses of bicyclic compounds having oxygen atoms have been done.

The X-ray[108] and electron diffraction[109] studies of bicyclo [2.2.1] heptane showed that it has a C_{2v} symmetry. The geometry may be approximated to that of 7-oxabicyclo [2.2.1] heptane V. The measured C–O–C angle in 7-oxabicyclo [2.2.1] heptane (94.5 ± 2.2°)[97, 110] is very close to the corresponding bridge angle in bicyclo [2.2.1] heptane (96°)[98]. As mentioned above, the ring strain energy of V is less than bicyclo [2.2.1] heptane due to the differences of the bond length and the bond repulsion. Incorporation of the two oxygens into the [2.2.1] systems may increase the reactivity because the oxygen atoms cause new repulsions between the unshared electron pairs. Thus, 2,6-dioxabicyclo [2.2.1] heptane VII and 2,7-dioxabicyclo [2.2.1] heptane VI are 6.9×10^5 and 2.5×10^4 times more reactive than their acyclic analogue, dimethylacetal[111]. On the other hand the hydrolysis rate constant for 2,6,7-trioxabicyclo [2.2.1] heptane VIII is *smaller* than those for triethyl and trimethyl orthoformates by a factor of 1.5 and 2, respectively[112]. The lower rate constant of VIII is explained by the occurrence of a different hydrolysis mechanism by Moreau et al.[113, 114]. Of course it is often difficult to correlate reactivity with strain energies or conformation.

Bicyclo [2.2.2] octane has the D_{3h} or D_3 symmetry (to lessen the repulsion of the eclipsing hydrogens at C_2 and C_3, there is a slight rotation around the three fold axis) and its strain energy is slightly less than that of bicyclo [2.2.1] heptane (Table 4). The geometry may be preserved in 2-oxa-(II), 2,6-dioxa-(III), and 2,6,7-trioxabicyclo [2.2.2] octane (IV). III showed high reactivity in the polymerization and the acid-hydrolysis

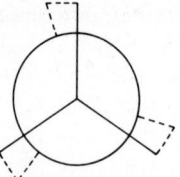

IX, D_3 symmetry in bicyclo [2.2.2] octane

(73.2 × 10^3 times more reactive than 1,1-dimethoxyethane)[54]. This high reactivity of III is explained in terms of the ring strain energy and Eliel's "rabbit-ear" effect[54]. III has a parallel lone pair situation. The lone pair interaction increases in IV, suggesting that IV has a high reactivity (in fact 2,6,7-trioxabicyclo [2.2.2] octanes show high polymerizability)[59].

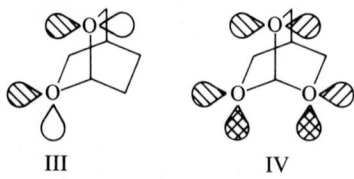

III   IV

Bicyclo [3.2.1] octane, in which a five-membered ring is fused to a cyclohexane ring at two axial positions, has C_s symmetry. The most stable conformation in cycloheptane is the "twist-chair", having a C_2 symmetry X[115, 116]. Even though the other conformations in cycloheptane are not so high in energy (the molecule readily undergoes pseudorotation), the cycloheptane ring in bicyclo [3.2.1] octane is locked in the "chair-boat" conformation.

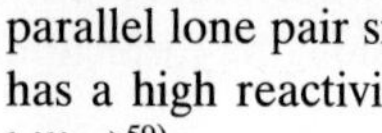

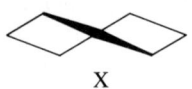

X

Some of the strain energy in bicyclo [3.2.1] octane is attributed to the nonbonded C_1–C_5 and/or the hydrogen repulsion at C_6 and C_7. 6,8-Dioxabicyclo [3.2.1] octane XI[42–49], 3,6,8-trioxabicyclo [3.2.1] octane XII[45], and 2,7,8-trioxabicyclo [3.2.1] octane XIII[58] show high polymerizability, indicating that they have comparable strain energies and Eliel's "rabbit-ear" effect as discussed for the bicyclic [2.2.2] systems.

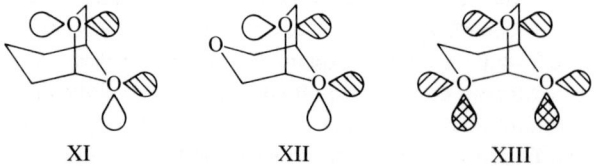

XI   XII   XIII

The most stable conformation of bicyclo [3.3.1] nonane is the "chair-chair" conformation[117–119]. The energy barriers between the double-chair and chair-boat and between the chair-boat and double-boat conformations are 27 and 43 kJ/mol, respectively[119]. The strain energy is less than for the [2.2.1], [3.2.1] and [2.2.2] systems (Table 4). Zefirov's group[120–124] has reported extensively on the conformational analysis of bicyclo [3.3.1] nonanes with hetero atoms at the 3-, 7-, and 9-positions. NMR studies using shift reagents, X-ray studies, photoelectron spectral data, and Hückel MO calculations showed that a double chair conformation is preferred in 3,7,9-trioxabicyclo [3.3.1] nonane XIV. 3-Oxa-XV and 3,7-dioxabicyclo [3.3.1] nonane XIV also exist in double-chair conformations, albeit with the wings slightly flattened because of lone pair repulsion[123]. On the contrary, the coupling constant, nuclear Overhauser effect, and T_1 relaxation times of 2,4-dioxabicyclo nonane XVII[125] show that it occurs predominantly in a boat-chair conformation, with the cyclohexane ring in a flattened chair form, contrary to previous reports by Anteunis et al.[126, 127]

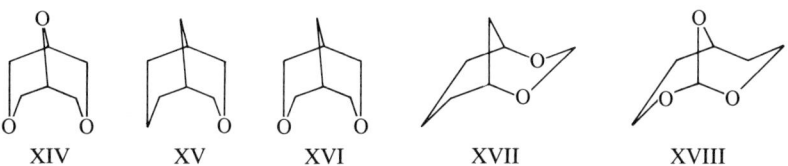

XIV   XV   XVI   XVII   XVIII

The boat-chair conformation may be due to the bond shortening at C_1–O_2 and O_4–C_5, i.e., the hydrogen repulsion at C_3 and C_7 increases in comparison with that in bicyclo [3.3.1] nonane. In this class, XV has no polymerizability[13], suggesting that the conformation of XV is a double chair and that the replacement of –CH_2– by O-atom at the –C_3– position lessens the strain energy.

On the other hand, XVIII readily undergoes polymerization to form oligomers and/or cyclic dimer[58]. ^{1}H-NMR of XVIII suggests that the conformation is not a double-chair, due to the p-orbital repulsion at O_2, O_8 and O_9 and the shortening at C_1–O_2 and C_1–O_8 bonds, i.e., the repulsion between the endo-hydrogens at C_3 and C_7 increases[59].

The conformation and the ring strain involved in bicyclo [3.3.0] octanes has been discussed by Armarego[102]. The cis-fused isomers are more stable than the trans ones. Therefore, trans-3-oxabicyclo [3.3.0] octane polymerizes but cis does not[30].

Anomeric Effect and Gauche Effect

The term "anomeric effect" was introduced by Lemieux[128] in 1958 in carbohydrate chemistry as a result of a detailed study of the conformational properties of acetylated pento- and hexopyranoses. The influence of lone-pair electrons is also discussed in the explanation of conformational properties in the molecules of X–C–Y, where X and Y are O, N, or another element having unshared electron lone pairs. The preference of the synclinal (gauche) conformation to the anti periplanar one is a characteristic phenomenon and has been generalized as the anomeric effect[9, 88, 129–132, 203]. Very recently Wolfe, Whangbo, and Mitchell[133] reviewed the "anomeric effect", "exo-anomeric effect" and "reverse-anomeric effect", and analyzed the C–X and C–Y bond lengths in X–CH_2–YH molecules. Interpretations of the anomeric effect have been given by many workers: dipole-dipole interaction between polar bonds[88, 131, 132, 134, 135], lone pair interactions[132, 136–141], and through-bond orbital interactions[135]. Anyway, the highest delocalization of an unshared electron pair of an oxygen atom occurs in a conformation in which the lone-pairs are situated antiperiplanar with respect to the C–Y bond: the synclinal (gauche) conformation for acyclic compounds and the axial one for heterocycles.

These favorable gauche states are forbidden in bicyclic compounds, i.e., at C_2–C_3 and C_5–C_6 for [2.2.1] system at C_2–C_3, C_5–C_6, and C_7–C_8 for [2.2.2] systems, and at C_6–C_7 for [3.2.1] systems. The unfavorable conformations are relieved by the ring-opening polymerization.

The geometry of bicyclo [2.2.1] heptane is C_{2v} symmetry[108, 109], and the dihedral angles are as follows:

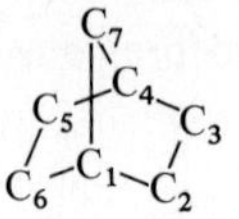

C_1–C_2–C_3–C_4 = C_4–C_5–C_6–C_1 planar

C_7–C_1–C_2–C_3 = C_7–C_1–C_6–C_5
C_7–C_4–C_3–C_2 = C_7–C_4–C_5–C_6 synperiplanar or synclinal

C_4–C_7–C_1–C_2 = C_4–C_7–C_1–C_6
C_1–C_7–C_4–C_3 = C_1–C_7–C_4–C_5 synclinal (gauche)

C_6–C_1–C_2–C_3 = C_3–C_4–C_5–C_6 synclinal

If the geometry of VI and VII is very similar to bicyclo [2.2.1] heptane, and if the gauche states are more important in the C–O↓C–O bond than in the C–C↓C–O bond, it may be concluded that VII is less stable than VI. In fact, the acid-hydrolysis data[112] show that VII is more reactive than VI. The difference of the hydrolysis rates is affected by the difference of the angles between C_1–C_7–C_4 and C_1–O_7–C_4 as well as the difference of Eliel's "rabbit-ear" effect between O_2–O_6 and O_2–O_7.

Eliel and Juaristi[142] reviewed the anomeric effect and gauche effect. Some of the results are shown here:

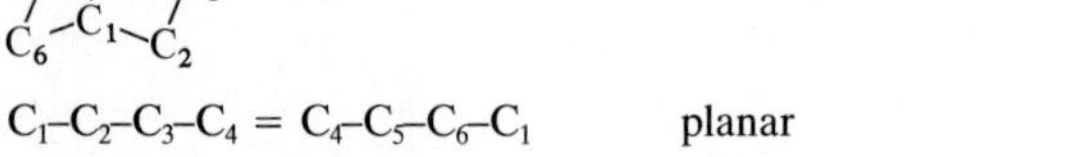

$\Delta G°$ = 3.1 kJ/mol in CCl_4[143]
1.5 kJ/mol in CH_3CN[143]

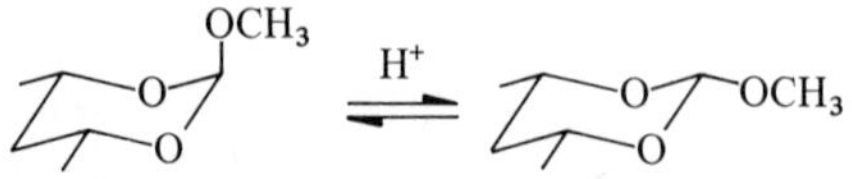

$\Delta G°$ = 1.7 kJ/mol in $C_2H_5OC_2H_5$[144]

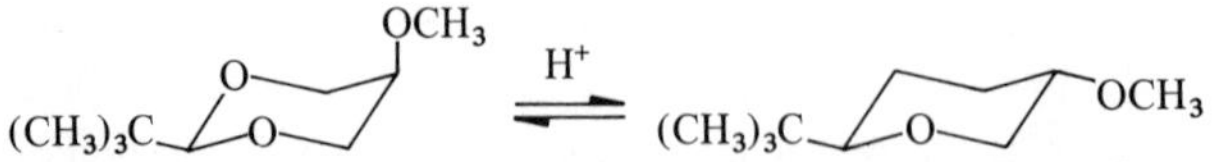

$\Delta G°$ = 0.67 kJ/mol in $CHCl_3$[145]
0.04 kJ/mol in CH_3CN[145]

These data give an indication of the magnitudes and directions of the conformational preference.

Electron diffraction, ab initio, and molecular mechanics studies in the gas phase[146] and the photoelectron spectra[147] of trimethyl orthoformate showed that it exists as an asymmetric TGG conformer in disagreement with the infrared spectroscopy study, which indicates a TGG$_T'$ conformation[148]. However, in bicyclic orthoesters all of the C–O bonds are locked in synclinal positions. The ring-opening relieves this unfavorable conformation.

4 Polymer Conformation

The characteristic preference of the synclinal (gauche) conformation over the trans one mentioned above has also been found for polymers containing oxygen ($[O(CH_2)_n]$) in both experimental and theoretical studies. In polyoxyethylene, gauche conformations are found to be ca. 6.3 kJ/mol lower in energy than the trans[149–152]. In polyoxyethylene, the gauche state is ca. 1.7 kJ/mol below the trans state[151, 153–156].

The gauche states are also observed in crystalline polymers, polyoxymethylene[6, 7, 157–160], polyoxyethylene[6, 7, 161, 162], poly-1,3-dioxolane[163], and poly-1,3-dioxepane[164] by X-ray analysis.

M. Mansson[165] measured the enthalpies of combustion and vaporization for some saturated, straight-chain oxa-compounds in the liquid and gaseous state at 25 °C; the next-nearest-neighbor oxygen atoms are shown to be stabler by 16.9 ± 0.2 kJ/mol per O–C–O interaction compared with monoethers. Snelson and Skinner[166] also pointed out the large increase in stability of 1,3-dioxane relative to 1,4-dioxane. Comparison between the ring strain energies (Table 4) and the thermodynamical data for the polymerization of five-membered monocyclic compounds[167] gives a very similar situation, i.e., the difference between the strain energy of tetrahydrofuran and 1,3-dioxolane is 2.5 kJ/mol but that for ΔH_{lc}^0 of the polymerization is 10 kJ/mol (The $-\Delta H_{lc}^0$ for the polymerization of cyclopentane, tetrahydrofuran, and 1,3-dioxolane is 50, 40 and 50 kJ/mol, respectively[167].). The difference in energy may in part be due to the high stabilization energy of the gauche state of polyoxymethylene in contrast to the trans state. The most favorable conformation of polyethylene is the planar zigzag trans form (TT···)[168]. As mentioned above, the gauche states are more favorable than the alternating trans states in polyoxymethylene (ca. 6.3 kJ/mol)[149–152] and polyoxyethylene (ca. 1.7 kJ/mol)[151, 153–156]. The

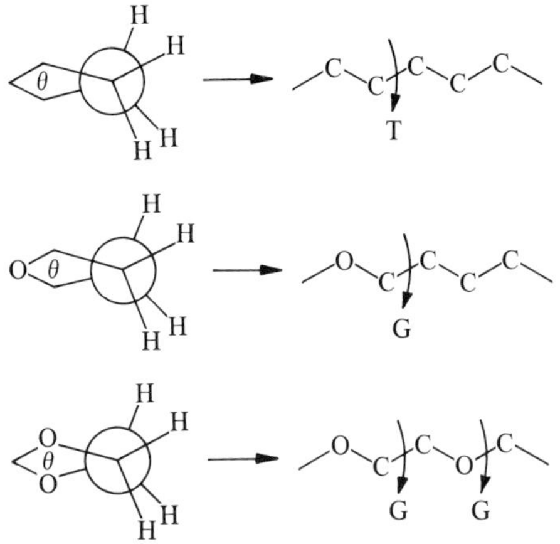

$\Theta < 60°$, synclinal or synperiplanar

stable conformation of polyoxytetramethylene is also the gauche state (ca. 0.8 kJ/mol)[169–171]. In cyclic compounds, those states are forbidden. This situation is not related to the planar, envelope, and twist forms conformations. The polymerizations relieve the unfavorable monomer conformations and their strains.

Polyethers

7-Oxabicyclo[2.2.1]heptane and its derivatives give polymers having a 1,4-trans-disubstituted cyclohexane ring[13, 20, 27]. The high stereoregular structure of poly-7-oxabicyclo-[2.2.1]heptane agrees with the high melting point of the polymer, 450 °C[13]. Saegusa et al.[20] studied the stereochemistry of the propagation of the cationic ring-opening polymerization of exo- and endo-2-methyl-7-oxabicyclo[2.2.1]heptane by means of NMR examination of the structure of the polymers. The endo-methyl monomer gave a polymer in which two ether groups as well as a methyl group are placed at the respective equatorial positions of a cyclohexane ring. In the polymer from the exo-methyl monomer, two ether groups on the cyclohexane ring are trans to each other and the methyl group is cis to the contiguous ether group. An equilibrium mixture of the two conformations XX and XXI has been assigned to the polymer of the exo-methyl monomer.

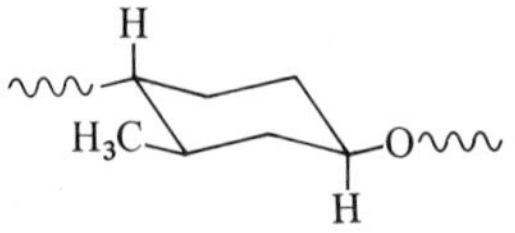

XIX
Poly-endo-2-methyl-7-oxabicyclo [2.2.1] heptane

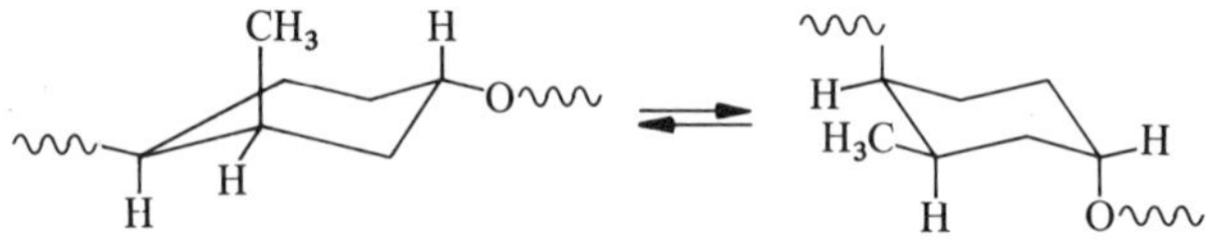

XX   XXI
Poly-exo-2-methyl-7-oxabicyclo [2.2.1] heptane

The equilibrium free energy difference between XX and XXI is assumed to be quite small because the energy difference values between the equatorial and axial conformers for the methyl (ΔG = 6.3–7.9 kJ/mol) and the two methoxy (ΔG = 2.1–2.9 kJ/mol) substituents of the cyclohexane ring are almost equal[64].

All bonds can be situated in gauche or trans states and the chain oxygens can be located opposite each other to relieve the orientation of the dipole moment vectors. The conformation may be similar to that of poly-tetrahydrofuran, planar zigzag[172, 173].

XXII

1,4-trans-Configuration for the cyclohexane ring is confirmed for the polymer from 2-oxabicyclo[2.2.2]octane II too[29].

Poly Acetals

Hall and coworkers[54] estimated the conformational equilibrium for both cis and trans isomers of poly-6,8-dioxabicyclo[3.2.1]octane and 2,6-dioxabicyclo[2.2.2]octane by the interplay of two factors: (1) the familiar preference of alkyl substituents to exist in the equatorial conformation and (2) the preference of the alkoxy group for the axial conformation (anomeric effect). The numerical parameters (kJ/mol) used for the calculations were[174, 175]: $OCH_{3ax} - H_{ax}$, 1.9; $CH_{3ax} - H_{ax}$, 3.8; $CH_{3ax} - OCH_{3ax}$, 10; $OCH_{3eq} -$ (anomeric effect), 5.4.

For poly-6,8-dioxabicyclo[3.2.1]octane[54].

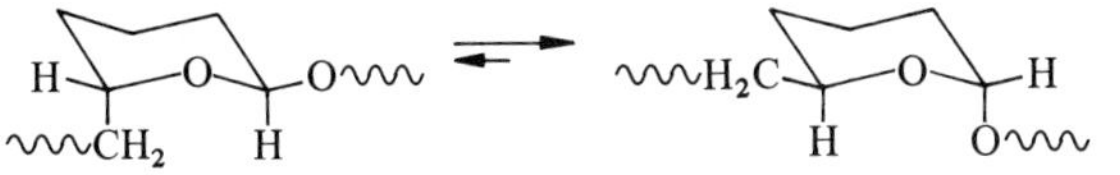

XXIII (1,3 ae) ⇌ XXIV (1,3 ea)
ΔG° = −9.2 kJ/mol
K = XXIV/XXIII = 42; XXIV = 98%

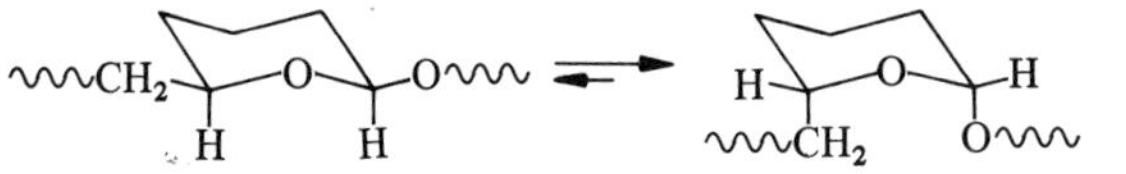

XXV (1,3 aa) ⇌ XXVI (1,3 ee)
ΔG° = −10.7 kJ/mol
K = XXVI/XXV = 81; XXV = 99%

Consequently, only the equatorial acetal hydrogen is observed experimentally for the polymer prepared at low temperature, indicating that trans 1,3-disubstituted tetrahydropyranoside is formed[42, 50].

The case for 2,6-dioxabicyclo[2.2.2]octane is not as simple because neither trans 1,4-units nor cis 1,4-units could be obtained conformationally homogeneous[54].

For poly-2,6-dioxabicyclo [2.2.2] octane[54].

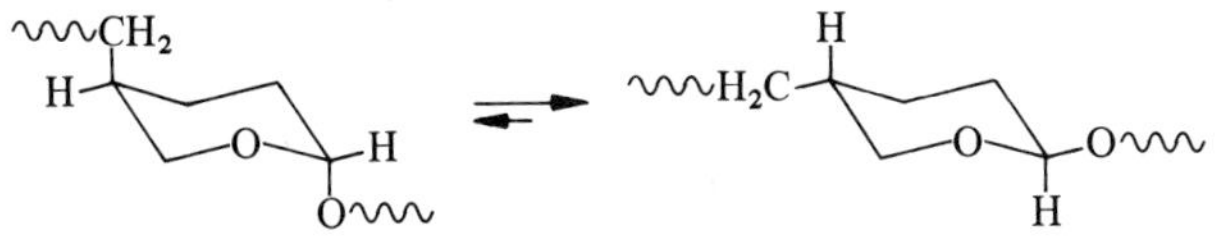

XXVII (1,4 aa) ⇌ XXVIII (1,4 ee)
ΔG° = −2.1 kJ/mol
K = XXVIII/XXVII = 2.3; XXVIII = 70%

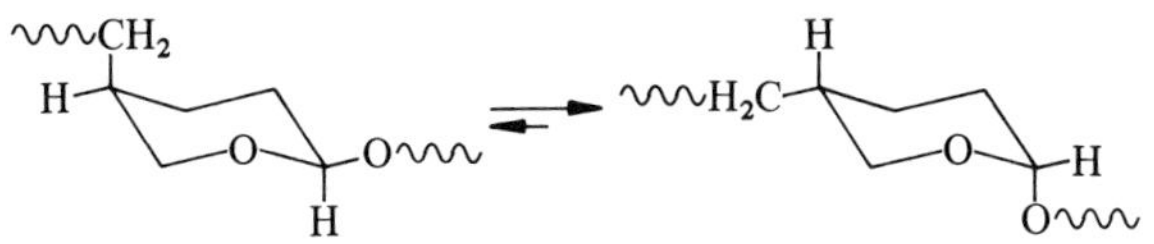

XXIX (1,4 ae) ⇌ XXX (1,4 ea)
ΔG° = −5.4 kJ/mol
K = XXX/XXIX = 9.1; XXX = 90%

The (H_{ax}/H_{eq}) ratios observed experimentally are, however, in good agreement with these calculated values.

2,6-Dioxabicyclo[2.2.1]heptane VII yields a polymer containing a 60:40 mixture of 2,4-disubstituted tetrahydro furan isomers (probably trans:cis) as detected by NMR[40]. 2,7-Dioxabicyclo[2.2.1]heptane VI gave a polymer containing exclusively tetrahydrofuran links at −78 °C, showing that a very high stereoselective cleavage at C_1–O_2 bond occurs. At high temperature, several initiators gave significant amounts of tetrahydropyran links[40]. Both polymers should be situated in a conformation with gauche and/or trans states for C–O and C–C bonds.

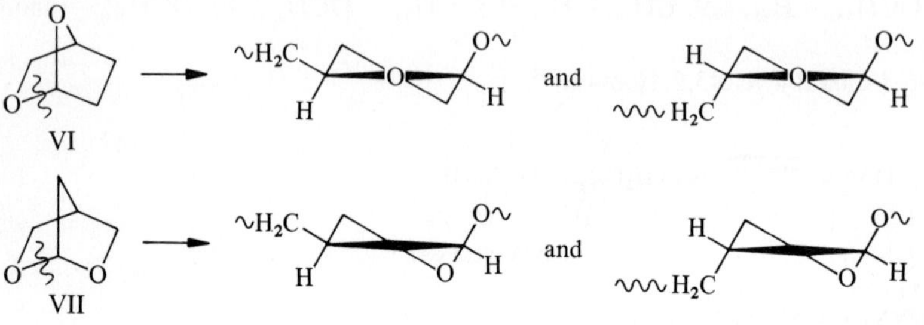

3,6,8-Trioxabicyclo[3.2.1]octane XII undergoes polymerization to give a polymer composed of cis and trans configurations for the tetrahydropyran links, even though the monomer structure is very similar to 6,8-dioxabicyclo[3.2.1]octane XI[45].

Poly-Orthoesters

2,6,7-Trioxabicyclo[2.2.1]heptane VIII[56] gives a pure poly-orthoester composed of cis and trans configurations for 1,3-dioxolane rings at low temperature. 2,7,8-Trioxabicyclo[3.2.1]octane XIII[58] gives a poly-orthoester having cis and trans configuration for the five-, six-, and seven-membered links even at low temperature. 2,6,7-Trioxabicyclo[2.2.2]octane IV and its derivatives[59] also give pure poly-orthoester at 0 °C.

In poly-VIII, the favorable conformation of synperiplanar-antiperiplanar-gauche for C–O bonds around the C_1-atom is not accomplished. The labile conformation of the polymer however, is more stable than that of the rigid monomer. In IV, all of the C–O bonds are situated in a gauche state. The polymer can exist as a TGG conformer, in which the position of the chain-methylene is determined by the exo-anomeric effect.

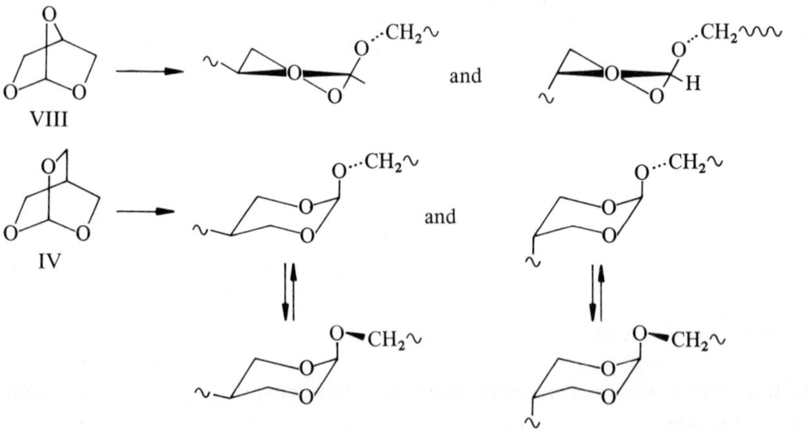

(1)

(2)

At high temperature two rings open to form poly-ethers with ester branching[60-63]. 2,6,7-Trioxabicyclo[2.2.1]heptane also gives a two ring-opened polymer at high temperature[57].

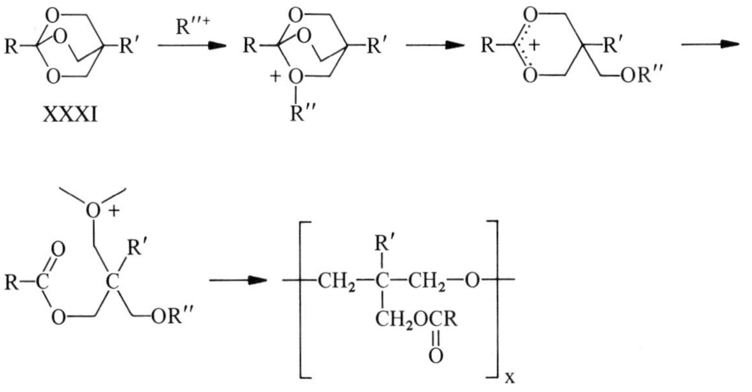

(3)

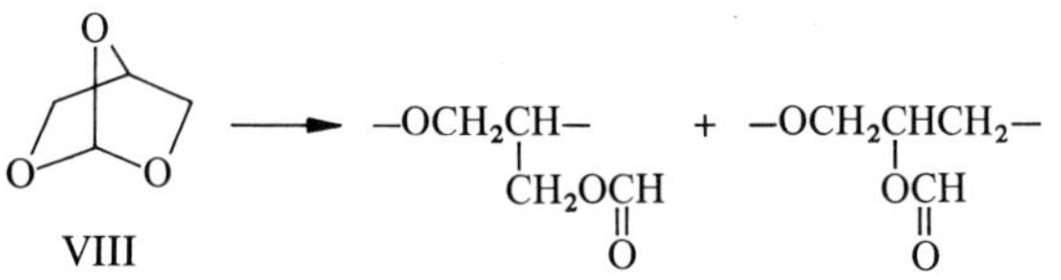

(4)

5 Kinetics

This subject has recently been authoritatively reviewed by Penczek, Kubisa, and Matyjaszewski[185].

For the kinetic study of the polymerization, the structure of the propagating species should be identified and the instantaneous concentration should be determined by chemical or physical method. Saegusa et al.[176-178] observed the propagating species in the polymerization of monocyclic ethers by ^{1}H NMR spectroscopy and used the "Phenoxyl End-Capping" method to determine the instantaneous concentration of the propagating species, Eq. (5). Both methods give the same concentrations.

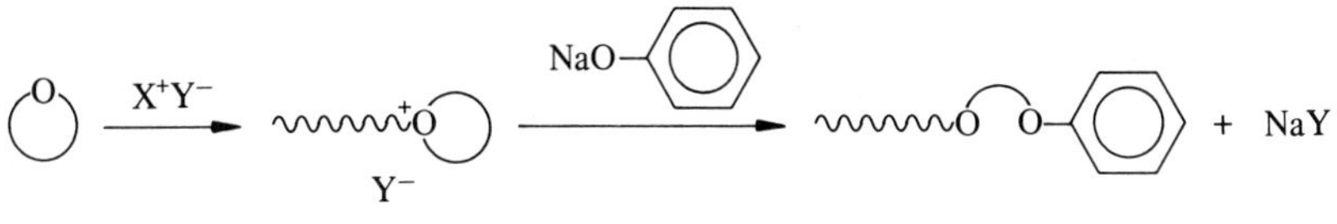

(5)

By using this method, they determined the rate constant of the propagation of endo- and exo-2-methyl-7-oxabicyclo[2.2.1]heptane[21, 22] and endo- and exo-2-tert-butyl-7-oxabicyclo[2.2.1]heptane[25]. In these systems the polymerization proceeds via S_N2 reaction between monomer and the propagating oxonium ion as was the case for the monocyclic ethers. They achieved the first case of a quantitative comparison of the polymerization reactivity between monocyclic and bicyclic ethers in the polymerization of exo-2-methyl-

7-oxabicyclo[2.2.1]heptane[21]: 1. The presence of the bridge increases the reactivity (k_p) and the ring strain of monomer. 2. Compared to the reactivity of THF the higher reactivity of exo-2-methyl-7-oxabicyclo[2.2.1]heptane is essentially ascribed to the large value of the frequency factor. In the polymerization of both 2-methyl- and 2-tert-butyl-7-oxabicyclo[2.2.1]heptanes, the endo-monomers show the higher reactivity. In the former case, the activation energy ($\Delta E_p^{\ddagger}$) plays the important role rather than the frequency factor, and in the latter case the frequency factor is more important. In another case endo, exo-2,6-dimethyl-7-oxabicyclo[2.2.1]heptane gives insoluble, highly crystalline polymers with phosphorous pentafluoride in dichloromethane, but exo, exo-monomer does not polymerize under identical conditions[27]; this again points to the higher reactivity of the endo-monomer.

Hvilsted and Kops[30] reported the kinetic study of trans-2- and trans-3-oxabicyclo[3.3.0]octane using the "Phenoxyl End-Capping" method. The initiation is more than three orders of magnitude slower than the propagation and the activation enthalpies are quite different for these reactions ($\Delta H_i^{\ddagger}$ = 74 and 71 kJ · mol^{-1} and $\Delta H_p^{\ddagger}$ = 62 and 56 kJ · mol^{-1} for the two monomers); this is being attributed to the relief of strain of the cyclic ion in the propagation. The rate constants for propagation are larger than those reported for THF and atom-bridged bicyclic substituted monomers of the 7-oxabicyclo[2.2.1]heptane series[21, 22, 25]. A mechanism is presented which includes the formation of two types of "dormant" tertiary oxonium ions involving the polymer chains. The authors[30] implied that it should be kept in mind that for these monomers, and probably in the case of other polymerizations involving the opening of an oxacyclopentane ring, many different active species may be present, including free ions and ion pairs.

In the polymerization of monocyclic ethers initiated with super acids such as methyl fluorosulfonate and methyl trifluoromethanesulfonate, two kinds of propagating species are observed in the reaction system and an extensive kinetic study has been done for the polymerization of THF, Eq. (6)[185]. However, the kinetic study of the polymerization of bicyclic ethers with super acid has not been reported.

$$\sim\sim\sim\overset{+}{\text{O}}\underset{Y^-}{\diagdown} \rightleftarrows \sim\sim\sim\text{O}(\text{CH}_2)_4-\text{Y} \qquad (6)$$

Y = OSO$_2$F, OSO$_2$CF$_3$, etc.

Table 5. Comparison of reactivities of mono- and bicyclic ethers

Monomer[a]	$10^3 k_p$ (at –20 °C) $\ell/(\text{mol} \cdot \text{s})$	$\Delta E_p^{\ddagger}$ kJ/mol	$10^6 A_p^{\ddagger}$ $\ell/(\text{mol} \cdot \text{s})$	$\Delta H_p^{\ddagger}$ kJ/mol	$\Delta S_p^{\ddagger}$ J/(mol · s)	Ref.
exo-MOBH	2.6	64.0	4 × 10^4	63.5	–63.5	21, 22
endo-MOBH	8.4	57.7	7 × 10^3	55.2	–41.8	22
exo-BOBH	2.5	46.4	5	42.6	–125	25
endo-BOBH	13	69.4	3 × 10^6	64.8	–25	25
trans-2-OBCO	56	63	10^6	63	–22	30
trans-3-OBCO	75	69	8 × 10^4	54	–46	30
THF	1.7 (at –10 °C)	50	11	–	–	21
THF (ion pair)	0.87 (at 0 °C)	–	–	–	–	179
THF (tree ion)	21 (at 0 °C)	–	–	–	–	179

[a] MOBH = 2-methyl-7-oxabicyclo[2.2.1]heptane, BOBH = 2-tert-butyl-7-oxabicyclo[2.2.1]heptane, and OBCO = oxabicyclo[3.3.1]octane

Moreover, the kinetic study of bicyclic acetals, bicyclic orthoesters, and other bicyclic monomers has not yet been done.

Kubisa and Penczek[181] reported that the Et_3O^+ cation is quantitatively transformed during initiation into Et_2O and a C_2H_5O-polymer end group (Eq. (6)) and that termination of the living poly-1,3-dioxolane-d_6 with Et_3N gave predominantly linear macromolecules of the following structure, $CH_3CH_2(OCD_2CD_2OCD_2)_nN^+(CH_3)_3$ (Eq. (7)) in the polymerization initiated with triethyloxonium hexafluoroantimonate.

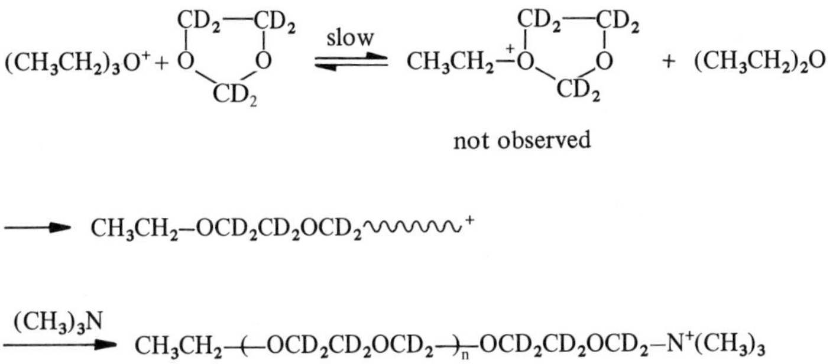

$$(CH_3CH_2)_3O^+ + \underset{CD_2}{\overset{CD_2-CD_2}{O\diagdown\diagup O}} \underset{\text{slow}}{\rightleftarrows} \underset{CD_2}{\overset{CD_2-CD_2}{CH_3CH_2-\overset{+}{O}\diagdown\diagup O}} + (CH_3CH_2)_2O$$

not observed

$$\longrightarrow CH_3CH_2-OCD_2CD_2OCD_2\sim\sim\sim\sim^+ \qquad (7)$$

$$\xrightarrow{(CH_3)_3N} CH_3CH_2\text{-(-}OCD_2CD_2OCD_2\text{-)}_n\text{-}OCD_2CD_2OCD_2\text{-}N^+(CH_3)_3$$

Triphenylphosphine has also been used to advantage as a trapping reagent in these polymerizations[185].

6 Polymerization Mechanism and Stereoselectivity

Bicyclic Ethers

The polymerization of bicyclic ethers occurs via the corresponding oxonium ion intermediates, resulting in one of two carbon atoms at the bridgeheads of the monomer being inverted during propagation as was shown in the polymerization of 7-oxabicyclo[2.2.1]heptanes[11]. The same situation is observed in the polymerization of 2-oxabicyclo[2.2.2]octane[29]. All the polymerization of bicyclic ethers listed in Table 1 suggest that the polymerizations proceed via S_N2 mechanisms, i.e., propagating species is a bicyclic oxonium ion.

Bicyclic Acetals

6,8-Dioxabicyclo[3.2.1]octane XI[42–49], and 2,6-dioxabicyclo[2.2.2]octane III[54], undergo polymerization to give highly stereoregular polymers at low temperature, suggesting an S_N2 propagation. The corresponding anion was observed in the polymerization of 1,6-anhydro-2,3,4-tri-O-benzyl-β-D-glucopyranose using PF_5 as a catalyst by means of ^{31}P and ^{19}F NMR spectroscopy[182]. The determination of kinetic isotope effects in the polymerization of 2,3,4-tri-O-methyllevoglucosan indicated that the polymerization active center is a cyclic oxonium ion[183].

Okada et al.[48, 49] studied the polymerization of optically active 6,8-dioxabicyclo[3.2.1]octane initiated with boron trifluoride etherate by means of ^1H and ^{13}C NMR spectroscopy, showing that the isotactic dyads, the D–D and L–L sequences, are formed preferentially. Conversion dependence of the specific rotation of the DL copolymers indicated that the L enantiomer, being in excess in the starting monomer mixture, was preferentially incorporated into the D, L copolymers chain. It was concluded that the preferred formation of the isotactic dyad sequence along the polymer chain at lower temperature is primarily due to the stereoregularity displayed by the growing chain end[49].

A less stereoregular propagation is observed for the polymerization of 6,8-dioxabicyclo[3.2.1]oct-3-ene, showing that the propagating chain end has some oxacarbenium ion character through the contribution of the neighboring carbon-carbon double bond[50–53].

In the polymerization of 2,6-dioxabicyclo[2.2.2]octane II, the most stereoregular polymer obtained at −78 °C using 0.2 mole % silicon tetrafluoride, possessed an (H_{ax}/H_{eq}) ratio of 3.0 compared with a calculated value of 2.3. This is regarded as indicating pure S_N2 propagation. The stereoregular propagation is influenced by temperature (most significant factor), the nature of the catalyst, the catalyst concentration, and trace impurities[54].

On the other hand, the stereoselective propagation of 2,7-dioxabicyclo[2.2.1]heptane VI and 2,6-dioxabicyclo[2.2.1]heptane VII[40] failed. At low temperature the polymer of VI contained exclusively tetrahydrofuran links; at high temperature several initiators gave significant amounts of tetrahydropyran links[40].

These results indicate that the polymerization mechanism of bicyclic acetals is as complicated as that of monocyclic acetals[184–188] and dependent on the monomer structure and the polymerization conditions.

Bicyclic Orthoesters

As mentioned above, poly-2,6,7-trioxabicyclo[2.2.1]heptane[56] and 2,7,8-trioxabicyclo[3.2.1]octane[58] contain the cis and trans configurations. The results suggest that the propagating chain ends have some oxacarbenium ion character, and as such the stereoregular propagation is poor. The contribution of oxacarbenium ion is due to the participation of the two adjacent oxygens.

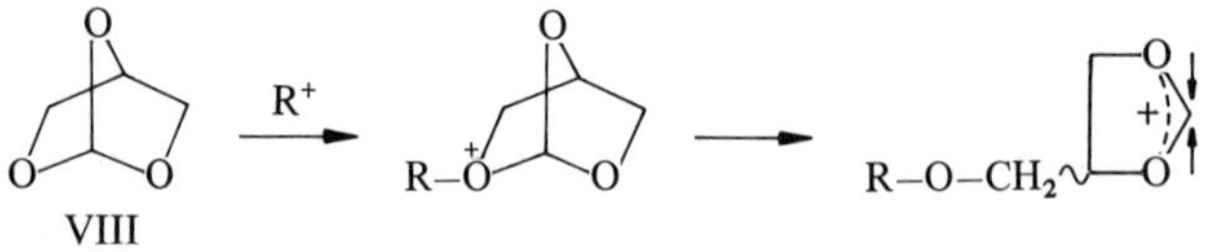

VIII

On the contrary, 2,6,7-trioxabicyclo[2.2.2]octanes XXXI[59] give polymers which are insoluble crystalline white powders with high melting points accompanied by decomposition. The NMR studies suggest that they possess the structure XXXII. The polymeriza-

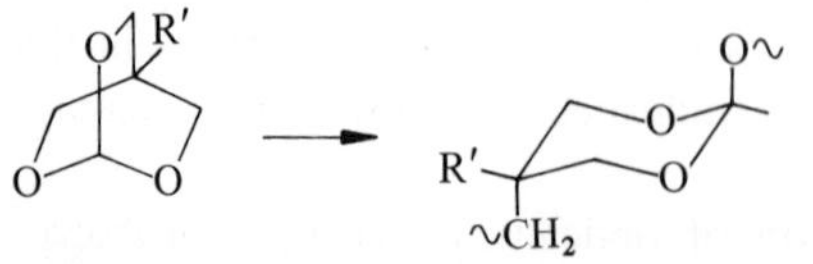

XXXI   XXXII

tion intermediate may be XXXIII. The weak dipole-dipole interaction between the 1,3-dioxolan-2-ylium ion and the chain oxygen atom contribute to the stereoregular propagation. The alternative oxonium ion XXXIV is not stable because four oxygen atoms are situated at the β-position of positive charge on an oxygen atom.

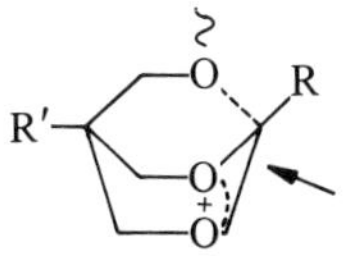

XXXIII 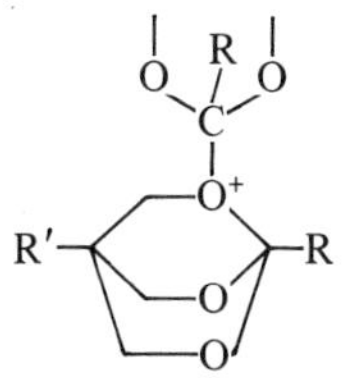   XXXIV

Generally, the acid-catalyzed hydrolysis of ortho esters proceeds through the following three steps: (1) generation of a dialkoxy carbonium ion, (2) hydration of the ion to form a hydrogen ortho ester, and (3) cleavage of the latter to form the hydrolysis products[187, 188]. Recently[189], UV spectroscopy at 25 °C has detected the 1,3-dioxolan-2-ylium ion and hydrogen ortho ester intermediates in the acid-catalyzed hydrolysis of a series of 2-aryl-(and 2-cyclopropyl)-2-alkoxy-1,3-dioxolanes[189]. Moreau et al.[113, 114] reported that in the hydrolysis of 2,4,10-trioxaadamantane and 3-methyl-2,4,10-trioxaadamantane the rate-determining step is the addition of water to the corresponding carboxonium ion intermediate. The oxygen adjacent to a carboxonium ion is a very strongly stabilizing group[190, 191]. These findings also suggest that the intermediate of the polymerization of bicyclic orthoesters are the corresponding 1,3-dioxolan- and 1,3-dioxan-2-ylium ions.

At high temperature, bicyclic orthoesters undergo polymerization to give poly-ethers with ester branching via two-ring-opening reactions[57, 60–63]. Poly-orthoesters are not isomerized to poly-ethers in dichloromethane with phosphorus pentafluoride and silicon tetrafluoride at –78, 0, and 22 °C. These results show that the formation of poly-ethers may be due to a decrease of the selective reactivity of the intermediate XXXV with the attacking monomer at higher temperatures. The formation of two kinds of polymer indicates that the intermediate is XXXV. The reaction between alkoxy-carbonium ions and nucleophiles was reviewed by Perst[192].

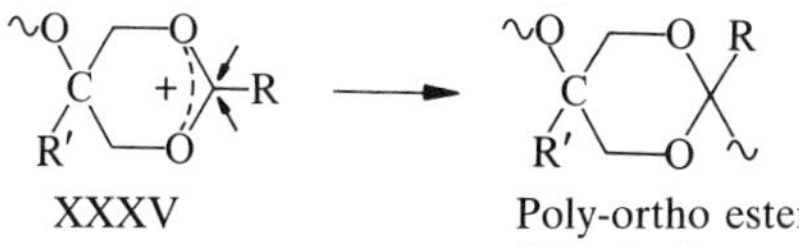

XXXV   Poly-ortho esters
            (kinetically controlled polymer)

XXXV   Poly-ethers
            (thermodynamically stable polymer)

2,6,7-Trioxabicyclo[2.2.1]heptane VIII gave five-membered units in the polymer chain[56, 57]. The selective formation of five-membered rings (or the selective cleavage at C_1–O_2 bond) indicates that one of intermediates is not reactive.

The stereoselective cleavage of a bond may be interpreted by the approach used by Deslongchamps et al.[193–195]. Specific cleavage of a carbon-oxygen or a carbon-nitrogen bond occurs when two heteroatoms (oxygen or nitrogen) of the tetrahedral intermediate each have one non-bonded electron pair oriented antiperiplanar to the departing O-alkyl

XXXVI

or N-alkyl group, XXXVI. This implies that the stereoselective cleavage in XXXVII$_a$ is a sufficiently low energy process. The alternative oxonium ion XXXVII$_b$ has no such situation. In XXXVIII$_a$, the situation is not complete but one p-orbital (shaded) is nicely

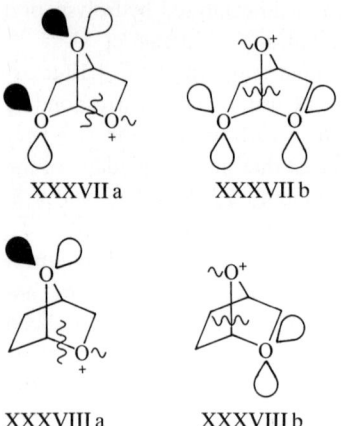

antiperiplanar to the C_1–O bond to be cleaved. Recently, Kirby and Martin[196, 197] reported that it is enough for such a specific cleavage of a bond that one p-orbital is oriented antiperiplanar to the departing group.

The results of a initio calculations on four selected conformations of $HOCH_2^+OH_2$ by Wipff[198] support the stereoselective cleavage. For the stereoselective cleavage, an extremely short lifetime of the intermediate is a very important parameter in relation to its reactivity.

VIII ⟶ XXXVII$_a$ and XXXVII$_b$ $\xrightarrow{\text{fast}}$ ~OCH$_2$-[structure]

As a result, the stereoregular polymerization of VI and VIII has not succeeded.

Form the above findings we can give a rough-hewn conclusion for the polymerization of bicyclic ethers, acetals, and ortho esters as is shown in Table 6.

Ring-Opening Polymerization 133

Table 6. Polymerization mechanism

Monomers	Mechanism	Stereoregular propagation
Bicyclic ethers	S_N2	+
Bicyclic acetals less hindered	S_N2 (at low temp.) S_N1 and S_N2 (at high temp.)	+ –
hindered	S_N1	–
Bicyclic ortho esters[a]	S_N1 (at low temp.)	–[b]

[a] At high temperature, two ring-opened polymer is formed
[b] Less hindered bicyclic orthoesters such as 2,6,7-trioxabicyclo[2.2.2]octanes may give stereoregular polymers

7 Medical Application of Polymers

Schuerch[2] reviewed the medical application of synthetic poly-saccharides. Capozza[199] reported that the poly-orthoester prepared from 2-ethoxy-4-hydroxymethyl-1,3-dioxolane in the presence of acid are useful as slow-release agents for drugs. Poly-2,6,7-trioxabicyclo[2.2.1]heptane VIII showed the same behavior[200].

Heller et al.[201, 202] showed that the polyacetals prepared by the condensation of a divinyl ether and a diol are only marginally useful for the release of norethindrone because hydrolysis of the acetal linkage at pH 7.4 is too slow. On the other hand the

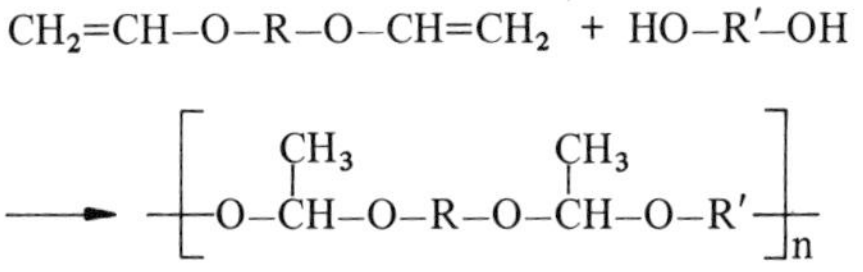

$$CH_2=CH-O-R-O-CH=CH_2 + HO-R'-OH$$

$$\longrightarrow \left[O-\underset{CH_3}{\overset{|}{CH}}-O-R-O-\underset{CH_3}{\overset{|}{CH}}-O-R' \right]_n$$

poly-orthoester produced from the following diketene acetal and 1,6-hexanediol exhibits a high drug release behavior at pH 7.4.

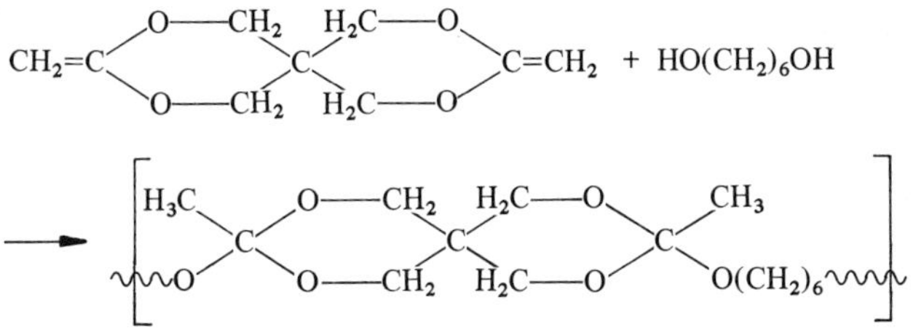

$$CH_2=C \overset{O-CH_2}{\underset{O-CH_2}{\diagup}} C \overset{H_2C-O}{\underset{H_2C-O}{\diagup}} C=CH_2 + HO(CH_2)_6OH$$

$$\longrightarrow \left[\underset{\sim\sim O}{\overset{H_3C}{\diagdown}} C \overset{O-CH_2}{\underset{O-CH_2}{\diagup}} \overset{H_2C-O}{\underset{H_2C-O}{\diagdown}} C \overset{CH_3}{\underset{O(CH_2)_6\sim\sim}{\diagup}} \right]$$

These results suggest that poly-orthoesters are also useful for medical applications. If hydroxyl or amino groups are incorporated into the skeleton, the corresponding polymers will become more valuable by being more water-soluble.

Acknowledgements. The authors are deeply indebted to the National Institutes of Health, Grant GM 18595, for support of this work.

8 References

1. Korshak, V. V. et al.: Vysokomol. Soedin. *3*, 477 (1961); CA *56*, 5830 (1962)
2. (a) Schuerch, C.: Adv. Polym. Sci. *10*, 173 (1972); (b) Schuerch, C.: Acc. Chem. Res. *6*, 184 (1973); (c) Schuerch, C.: Encycl. Polym. Sci. and Technol. Suppl., Vol. 1, 510 (1976)
3. Kobayoshi, S., Eby, R., Schuerch, C.: Biopolymers *16*, 415 (1977)
4. Sumitomo, H., Okada, M.: Adv. Polym. Sci. *28*, 47 (1978)
5. Tadokoro, H.: Macromol. Rev. *1*, 119 (1967)
6. Tadokoro, H.: Structure of Crystalline Polymers, Wiley-Interscience, New York, 1979
7. Mark, J. E.: Acc. Chem. Res. *12*, 49 (1979)
8. Eliel, E. L. et al.: Conformation Analysis, Wiley-Interscience, New York, N. Y., 1965
9. Szarek, W. A., Horton, D., (Ed.): Anomeric Effect, Origin and Consequences, ACS Symp. Series 87, Amer. Chem. Soc., Washington, D. C., 1979
10. (a) Bailey, W. J.: J. Macromol. Sci. Chem. *A 9*, 849 (1975); (b) Bailey, W. J., Iwama, H., Tsushima, R.: J. Polym. Sci., Symp. *56*, 117 (1976); (c) Endo, T., Bailey, W. J.: Makromol. Chem. *176*, 2897 (1975); (d) Endo, T., Katsuki, H., Bailey, W. J.: ibid. *177*, 3231 (1970)
11. Wilkins, J. P.: U.S. Pat. 2, 764, 559 (1956); CA *51*, 4053g (1957)
12. Campbell, T. W.: U.S. Pat. 2, 831, 825 (1958); CA *52*, 13316i (1958)
13. Wittbecker, E. L., Hall, H. K. Jr., Campbell, T. W.: J. Amer. Chem. Soc. *82*, 1218 (1960)
14. Giusti, P., Andruzzi, F.: Ann. Chem. (Rome) *56*, 973 (1966)
15. Magagnini, P. L. et al.: Ann Chem. (Rome) *57*, 1493 (1967)
16. Baccaredda, M. et al.: Chim. Ind. (Milano) *50*, 81 (1968)
17. Giusti, P., Andruzzi, F., di Maina, M.: Chin. Ind. (Milano) *50*, 121 (1968)
18. Giusti, P. et al.: Makromol. Chem. *128*, 1 (1969)
19. Baccaredda, M. et al.: J. Polymer Sci. Part C *31*, 157 (1979)
20. Saegusa, T. et al.: Macromolecules *5*, 233 (1972)
21. Saegusa, T. et al.: ibid. *5*, 236 (1972)
22. Saegusa, T. et al.: ibid. *5*, 815 (1972)
23. Andruzzi, F., Barnes, D. S., Plesch, P. H.: Makromol. Chem. *176*, 2053 (1975)
24. Saegusa, T., Motoi, M., Suda, H.: Macromolecules *9*, 231 (1976)
25. Saegusa, T., Motoi, M., Suda, H.: ibid. *9*, 526 (1976)
26. Andruzzi, F. et al.: Makromol. Chem. *178*, 2367 (1977)
27. Kops, J., Spanggaard, H.: Makromol. Chem. *151*, 21 (1972)
28. Saegusa, T. et al.: Polym. J. *11*, 113 (1979)
29. Saegusa, T., Hodaka, T., Fujii, H.: Polym. J. *2*, 670 (1971)
30. Hvilsted, S., Kops, J.: Macromolecules *12*, 889 (1979)
31. Kops, J., Larsen, E., Spanggaard, H.: J. Polym. Sci., Polym. Symp. *56*, 91 (1976)
32. Kops, J., Spanggaard, H.: Makromol. Chem. *175*, 3077 (1974)
33. Crivello, J. V., Lam, J. H. W.: J. Polym. Sci., Polym. Symp. *56*, 383 (1976)
34. Chang, E. Y. C.: Ph. D. Dissertation, Univers. Pennsylvania, 1961
35. Ceccarelli, G., Andruzzi, F.: Makromol. Chem. *180*, 1371 (1979)
36. Irie, M., Yamamoto, Y., Hayashi, K.: Pure Appl. Chem. *49*, 455 (1977)
37. Crivello, J. V., Lam, J. H. W., Volantee, C. N.: J. Radiat. Curing *4*, 2 (1977)
38. Malhotra, S. L., Blanchard, L. P.: J. Macromol. Sci. Chem. *A 13*, 1379 (1978)
39. Irie, M. et al.: J. Polym. Sci., Polym. Chem., *17*, 815 (1979)
40. Hall, H. K. Jr. et al.: J. Polym. Sci., Polym. Symp. *56*, 101 (1976)
41. Okada, M., Sumitomo, H., Irii, S.: Makromol. Chem. *177*, 2331 (1976)
42. Tamura, A. et al.: Kogyo Kagaku Zasshi (J. Chem. Soc. Jpn., Ind. Chem. Sect.) *68*, 2271 (1965)
43. Kops, J.: J. Polym. Sci. Part A 1, *10*, 1275 (1972)
44. Sumitomo, H., Okada, M., Hibino, Y.: J. Poly. Sci., Polym. Lett. Ed. *10*, 871 (1972)
45. Hall, H. K. Jr., Steuck, M. J.: J. Polym. Sci., Polym. Chem. Ed. *11*, 103 (1973)
46. Okada, M., Sumitomo, H., Hibino, Y.: Polym. J. *6*, 256 (1974)
47. Okada, M., Sumitomo, H., Hibino, Y.: ibid. *7*, 511 (1975)
48. Komada, H., Okada, M., Sumitomo, H.: Macromolecules *12*, 5 (1979)

49. Okada, M., Sumitomo, H., Komada, H.: ibid. *12*, 395 (1979)
50. Okada, M., Sumitomo, H., Komada, H.: ibid. *178*, 343 (1977)
51. Okada, M., Sumitomo, H., Komada, H.: ibid. *179*, 949 (1978)
52. Komada, H., Okada, M., Sumitomo, H.: ibid. *179*, 2859 (1978)
53. Okada, M. et al.: ibid. *180*, 813 (1979)
54. Hall, H. K. Jr. et al.: J. Amer. Chem. Soc. *96*, 7265 (1974)
55. Kops, J., Spanggaard, H.: Makromol. Chem. *176*, 299 (1975)
56. Yokoyama, Y. et al.: Macromolecules *13*, 252 (1980)
57. Hall, H. K. Jr., Yokoyama, Y.: Polym. Bull. *2*, 281 (1980)
58. Yokoyama, Y., Hall, H. K. Jr.: J. Polym. Sci., Polym. Chem. Ed. *18*, 3133 (1980)
59. Yokoyama, Y. et al.: Macromolecules, submitted
60. Bailey, W. J. et al.: ACS Symp. Series *59*, 38 (1977)
61. Bailey, W. J., Saigo, K.: Abstr. ACS Meet. Houston, March, 1980, p. 4
62. Endo, T., Saigo, K., Bailey, W. J.: J. Polym. Sci., Polym. Lett. *18*, 457 (1980)
63. Endo, T. et al.: J. Poly. Sci., Polym. Lett. *18*, 771 (1980)
64. Goethals, E. J.: Adv. Polym. Sci. *23*, 103 (1977)
65. Yamashita, Y., Kawakami, Y., in: Ring-Opening Polymerization, (Ed. T. Saegusa, E. Goethals) Amer. Chem. Soc., Washington, D. C., 1977, p. 99
66. Smis, D.: J. Chem. Soc. *1964*, 864
67. Dreyfuss, M. P., Dreyfuss, P.: J. Polym. Sci., Part A-1 *4*, 2179 (1966)
68. Plesch, P. H., Westermann, P. H.: J. Polym. Sci., Part C *16*, 3837 (1968)
69. Yamashita, Y. et al.: Makromol. chem. *114*, 146 (1968)
70. Hall, H. K. Jr.: J. Amer. Chem. Soc. *80*, 6412 (1958)
71. Dainton, F. S., Ivin, K. J.: Quart. Rev. (London) *12*, 61 (1968)
72. Saegusa, T. et al.: Polym. J. *3*, 40 (1972)
73. Small, P. A.: Trans. Faraday Soc. *51*, 1717 (1955)
74. Okada, M., Mita, K., Sumitomo, H.: Makromol. Chem. *176*, 859 (1975)
75. Okada, M., Mita, K., Sumitomo, H.: ibid. *177*, 2055 (1976)
76. Cox, J. D.: Tetrahedron *19*, 1175 (1963)
77. Greenberg, A, Liebman, J. F.: Strained Organic Molecules, Academic Press, New York, San Francisco and London, 1978
78. Benson, S. W.: Thermochemical Kinetics, John-Wiley & Sons, New York, 1976, p. 60
79. Ref. (26); these data were derived from the Allen bond-energy scheme [Allen, T. L.: J. Chem. Phys. *31*, 1039 (L 959)]
80. Cox, J. D., Pilcher, G.: Thermochemistry of Organic and Organometallic Compounds, Academic Press, New York and London, 1970
81. Greenhouse, J. A., Strauss, H. L.: J. Chem. Phys. *50*, 124 (1969)
82. Engerholm, G. G. et al.: J. Chem. Phys. *50*, 2446 (1969)
83. Cremer, D., Pople, J. A.: J. Amer. Chem. Soc. *97*, 1358 (1975)
84. Lemieux, R. U., Stevens, J. D., Fraser, R. R.: Can. J. Chem. *40*, 1955 (1962)
85. Gratti, G., Segre, A. L., Morandi, C.: J. Chem. Soc. (13) *1967*, 1203
86. Romers, C. et al.: Topics Stereochem. *4*, 38 (1969)
87. Riddell, F. G.: Quart. Rev. *21*, 364 (1967)
88. Eliel, E. L.: Angew. Chem., Internat. Edn. *11*, 739 (1972)
89. De Kok, A. J., Romers, C.: Rec. Trav. Chim. *89*, 313 (1970)
90. Nader, F. W.: Tetrahedron Lett. *1975*, 1207
91. Nader, F. W.: ibid. *1975*, 1591
92. Barbier, C., Delman, J., Rault, J.: ibid. *1964*, 3339
93. Anteunis, M., Tavernier, D., Borremans, F.: Bull. Soc. Chim. Belges *75*, 396 (1966)
94. Pihlaja, K., Heikkila, J.: Acta Chem. Scand. *21*, 2430 (1967)
95. Lambert, J. B.: Acc. Chem. Res. *4*, 87 (1971)
96. Forrest, T. P.: J. Amer. Chem. Soc. *97*, 2628 (1975)
97. Oyanagi, K. et al.: Bull. Chem. Soc. Japan *48*, 751 (1975)
98. Chiang, J. F., Wilcox, C. F. Jr., Bauer, S. H.: J. Amer. Chem. Soc. *90*, 3149 (1968)
99. Chiurdoglu, G., (Ed.): Conformational Analysis, Academic Press, New York, London, 1971
100. Hanack, M.: Conformational Theory, Academic Press, New York, 1963

101. McKenna, J.: Conformational Analysis of Organic Compounds, Royal Inst. Chem., Lecture Ser. London, No. 1, 1966
102. Armarego, W. L. F., Gallagher, M. J.: Stereochemistry of Heterocyclic Compounds (Taylor, E. C., Weissberger, A., Ed.), Wiley-Interscience, New York, 1977
103. Packer, W., (Ed.): Saturated Heterocyclic Chemistry, Vol. 1, Chem. Soc., London, 1973
104. Packer, W., (Ed.): ibid. Vol. 2, Chem. Soc., London, 1974
105. Ansell, M. F., (Ed.): ibid. Vol. 3, Chem. Soc., London, 1975
106. Ansell, M. F., Pattenden, G., (Ed.): ibid. Vol. 4, Chem. Soc., London, 1977
107. Pattenden, G., (Ed.): ibid. Vol. 5, Chem. Soc., London, 1978
108. Altona, C., Sundaralingam, M.: J. Amer. Chem. Soc. *92*, 1995 (1970)
109. Dallinga, G., Toneman, L. H.: Rec. Trav. Chim. *87*, 795 (1968)
110. Creswell, R. A.: J. Mol. Spectrosc. *56*, 133 (1975)
111. Hall, H. K. Jr., De Blauwe, Fr.: J. Amer. Chem. Soc. *97*, 655 (1975)
112. Hall, H. K. Jr., De Blauwe, Fr., Pyriadi, T.: ibid. *97*, 3854 (1975)
113. Bouab, O., Moreau, C., Zeh. Ako, M.: Tetrahedron Lett. *1978*, 61
114. Bouab, O., Lamaty, G., Moreau, C.: J. Chem. Soc., Chem. Comm. *1978*, 678
115. Hendrickson, J. B. et al.: J. Amer. Chem. Soc. *95*, 494 (1973)
116. Glazer, E. S. et al.: ibid. *94*, 6026 (1972)
117. Engler, E. M., Andose, J. D., Schleyer, P. von R.: ibid. *95*, 8005 (1973)
118. Peters, J. A. et al.: Tetrahedron *34*, 3313 (1978)
119. Mikhailov, V. K. et al.: Izv. Akad. Nauk. SSSR, Ser. Khim. *11*, 2455 (1978); CA *90*, 86625 a (1979)
120. Zefirov, N. S., Kurkutova, E. N., Gonchurov, A. V.: Zhur. Org. Khim. *10*, 1124 (1974)
121. Goncharov, A. V. et al.: Dokl. Akad. Nauk. SSSR *214*, 505 and 810 (1974)
122. Zefirov, N. S. et al.: J. Chem. Soc. Chem. Comm. *1974*, 260
123. Zefirov, N. S., Rogozina, S. V.: Tetrahedron *30*, 2345 (1974)
124. Gleiter, R. et al.: Dokl. Akad. Nauk, SSSR *235*, 347 (1977)
125. Peters, J. A. et al.: Tetradedron Lett. *1979*, 2553
126. Swaelens, G., Anteunis, M.: Bull. Soc. Chim. Belges *78*, 471 (1969)
127. Anteunis, M., Bécu, C., Anteunis-ole Ketelaere, F.: J. Acta. Crencia Inica *1*, 1 (1974)
128. Lemieux, R. U., Chu, N. J.: Abstr. Amer. Chem. Soc. *133*, 31 N (1958)
129. Jeffrey, G. A. et al.: J. Amer. Chem. Soc. *100*, 373 (1978)
130. Tvaroska, I., Bleha, T.: Can. J. Chem. *57*, 424 (1979)
131. Lemieux, R. U.: Pure & Appl. Chem. *25*, 527 (1971)
132. De Houg, A. J. et al.: Tetradedron *25*, 3365 (1969)
133. Wolfe, S., Whangbo, M.-H., Mitchell, D.: Carbohydr. Res. *69*, 1 (1979)
134. Edward, J. T.: Chem. Ind. (London) *1955*, 1102
135. Traroška, I., Bleha, T.: Can. J. Chem. *57*, 424 (1979)
136. Lucken, E. A. C.: J. Chem. Soc. *1959*, 2954
137. Romers, C. et al.: Top. Stereochem. *4*, 39 (1969)
138. Baddeley, G.: Tetrahedron Lett. *1973*, 1645
139. David, S. et al.: J. Amer. Chem. Soc. *95*, 3806 (1973)
140. Bürg, H. B. et al.: Tetrahedron *30*, 1563 (1974)
141. Jeffrey, G. A., Pople, J. A., Radom, L.: Carbohydr. Res. *25*, 117 (1972)
142. Eliel, E. L., Juaristi, E.: in Ref. (8), pp. 95–106
143. Eliel, E. L., Giza, C. A.: J. Org. Chem. *33*, 3754 (1968)
144. Nader, F. W., Eliel, E. L.: J. Amer. Chem. Soc. *92*, 3050 (1970)
145. Kaloustian, M. K. et al.: J. Amer. Chem. Soc. *98*, 956 (1976)
146. Spelbos, A., Mijlhoff, F. C., Faber, D. H.: J. Mol. Struct. *4*, 47 (1977)
147. Gan, T. H., Peel, J. B., Willett, G. D.: Chem. Phys. Lett. *51*, 464 (1977)
148. Lee, H., Wilmshurst, J. K.: Spectrochim. Acta. *23 A*, 247 (1967)
149. Uchida, T., Kurita, Y., Kubo, M.: J. Polym. Sci. *19*, 365 (1956)
150. Flory, P. J., Mark, J. E.: Makromol. Chem. *75*, 11 (1964)
151. Flory, P. J.: Statistical Mechanics of Chain Molecules, Interscience, New York, 1969
152. Mark, J. E.: Acc. Chem. Res. *7*, 218 (1974)
153. Mark, J. E., Flory, P. J.: J. Amer. Chem. Soc. *87*, 1415 (1965)

154. Mark, J. E., Flory, P. J.: ibid. 88, 3702 (1966)
155. Bak, K., Elefante, G., Mark, J. E.: J. Phys. Chem. 71, 4007 (1967)
156. Patterson, G. D., Flory, P. J.: J. Chem. Soc., Faraday Trans. 2 68, 1111 (1972)
157. Tadokoro, H. et al.: J. Polym. Sci. 44, 266 (1960)
158. Garazzolo, G.: ibid. A, 1, 1573 (1963)
159. Garazzolo, G., Mammi, M.: ibid. A, 1, 965 (1963)
160. Uchida, T., Tadokoro, H.: ibid. A-2, 5, 63 (1967)
161. Takahashi, Y., Tadokoro, H.: Macromolecules 6, 672 (1973)
162. Takahashi, Y., Sumita, I., Tadokoro, H.: J. Polym. Sci., Polym. Phys. Ed. 11, 2113 (1973)
163. Sasaki, S., Takahashi, Y., Tadokoro, H.: ibid 10, 2363 (1972)
164. Sasaki, S., Takahashi, Y., Tadokoro, H.: Polym. J. 4, 172 (1973)
165. Mansson, M.: J. Chem. Thermodynamics 1, 141 (1969)
166. Snelson, A., Skinner, H. A.: Trans. Faraday Soc. 57, 2125 (1961)
167. Busfield, W. K., Lee, R. M., Merigold, D.: Makromol. Chem. 156, 183 (1972)
168. Bunn, C. W.: Trans. Faraday Soc. 35, 482 (1939)
169. Mark, J. E.: J. Polym. Sci., Part B 4, 825 (1966)
170. Mark, J. E.: J. Amer. Chem. Soc. 88, 3708 (1966)
171. Abe, A., Mark, J. E.: J. Amer. Chem. Soc. 98, 6468 (1976)
172. Imada, K. et al.: Makromol. Chem. 83, 113 (1965)
173. Cesari, M., Perego, G., Mazzei, A.: ibid. 83, 196 (1965)
174. Angyal, S. K.: Austr. J. Chem. 21, 2737 (1968)
175. Anderson, C. B., Sepp, D. T.: Tetrahedron 24, 1707 (1968)
176. Saegusa, T., Matsumoto, S.: Macromolecules 1, 442 (1968)
177. Saegusa, T., Matsumoto, S.: J. Polym. Sci. Part A-1 6, 1559 (1968)
178. Saegusa, T., Matsumoto, S., Hashimoto, Y.: Macromolecules 4, 1 (1971)
179. Bourdauduca, P., Worsfold, D. J.: ibid. 8, 562 (1975)
180. For example, (a) Saegusa, T., Kobayashi, S.: J. Polym. Sci., Polym. Symp. 56, 241 (1976); (b) Penczek, S., Matyjaszewski, K.: J. Polym. Symp. 56, 255 (1976)
181. Kubisa, P., Penczek, S.: Macromolecules 10, 1216 (1977)
182. Uryu, T. et al.: Makromol. Chem. 180, 1509 (1979)
183. Sakharov, A. M. et al.: Dokl. Akad. Nauk. SSSR 250, 1381 (1980); CA 92, 147287 g (1980)
184. Plesch, P. H.: Pure & Appl. Chem. 48, 287 (1967)
185. Penczek, S., Kubisa, P., Matyjaszewski, K.: Adv. Polym. Sci. 37, 1 (1980)
186. (a) Yamashita, Y., Kawakami, Y., Kitano, K.: J. Polym. Sci., Polym. Lett. B 15, 213 (1977); (b) Kawakami, Y., Suzuki, J., Yamashita, Y.: Polym. J. 9, 519 (1977); (c) Kawakami, Y., Yamashita, Y.: Macromolecules 12, 399 (1979); They reported that the polymerization of 1,3,6,9-tetraoxacycloundecane proceeds in two stages, in the first cyclic oligomers are formed via an oxonium ion mechanism and in the second stage high polymers are mainly formed via a carbocation mechanism
187. Cordes, E. H., Bull, H. G.: Chem. Rev. 74, 581 (1974)
188. Fife, T. H.: Acc. Chem. Res. 8, 264 (1972)
189. Ahmad, M. et al.: J. Amer. Chem. Soc. 101, 2669 (1979)
190. Ramsy, B. C., Taft, R. W.: ibid 88, 3058 (1966)
191. Taft, R. W., Martin, R. H., Lampe, F. W.: ibid 87, 2490 (1965)
192. Perst, H.: Oxonium Ions in Organic Chemistry, Academic Press, New York, 1971
193. Deslongchamps, P.: Tetrahedron 31, 2463 (1975)
194. Deslongchamps, P. et al.: Can. J. chem. 53, 1601 (1975)
195. Deslongchamps, P.: Heterocycles 7, 1271 (1977)
196. Kirby, A. J., Martin, R. J.: Chem. Commun. 1978, 803
197. Kirby, A. J., Martin, R. J.: ibid. 1979, 1079
198. Wipff, G.: Tetrahedron Lett. 1978, 3269
199. Capozza, R. C., Otten, Gu.: U.S. Pat. 2, 715, 502, Oct., 1977; CA 88, 51358 x (1977)
200. Hall, H. K. Jr., Yokoyama, Y.: unpublished
201. Heller, J., Penhale, D. W. H., Helwing, R. F.: J. Polym. Sci., Polym. Lett. 18, 293 (1980)
202. Heller, J., Penhale, D. W. H., Helwing, R. F.: Polym. Preprints 21, No. 2, p. 82, Las Vegas, August, 1980

203. Tvaroska, I., Bleha, T.: Collect. Czech Chem. Commun. *45,* 1883 (1980); CA *94,*15031a (1981); They studied the stability of the $(MeO)_2CH_2$ conformer formed by rotation about the central C–O bonds. The anomeric (gauche) effect makes the synclinal more stable than the antiperiplanar conformations. The anomeric effect disappears with increasing solvent polarity. Their results suggest that the anomeric effect still contributes to the stability of the monomer and polymer conformation in polymerization solvents such as dichloromethane

Received May 5, 1981
J. K. Stille (editor)

Structural and Chemical Modifications of Cellulose by Graft Copolymerization

B. P. Morin, I. P. Breusova, and Z. A. Rogovin

Moscow Textile Institute, M. Kaluskaja I, Moscow B-419, USSR

Zakhar Aleksandrovich Rogovin

*28. 8. 1905   †11. 8. 1981

Studies concerning the following problems are reviewed;
1. Influence of the cellulose structure and of the chemical structure of the grafted polymer on the kinetics of graft copolymerization of cellulose, on the composition, and the structure of the cellulose products.
2. Kinetics of chemically initiated graft polymerization onto cellulose and its peculiarities connected with the heterogeneity of the process.
3. Specificities of grafting in the heterogenious medium and use of some effects in technology.
4. Basic trends in the development of the commercial process of graft polymerization on cellulose.
5. Polymer-analoguous conversion of polymer chains grafted to cellulose.

List of Symbols 140

1 Introduction 141

2 Factors Influencing Graft Copolymerization with Cellulose, Structure and Properties of its Products 141
   2.1 Effect of Cellulose Structure and Surface Area on the Kinetics of Grafting and Structure of the Products of Grafting 141
   2.2 Influence of the Cellulose Macroradical Accessibility on the Kinetics and Products of Graft Polymerization 143
   2.3 Grafting on Low-Substituted Cellulose Derivatives 144
   2.4 Cellulose Conversion During Graft Polymerization 144
   2.5 Influence of the Cellulose Matrix on the Orientation of the Grafted Chains 146
   2.6 Grafting on Cross-Linked Cellulose 146
   2.7 Morphology of the Products of Cellulose Grafting 148
   2.8 Influence of the Grafted Polymer Distribution on the Properties of the Grafted Copolymers 150

3 Kinetics of Chemically Initiated Graft Copolymerization and its Features Arising from the Heterogeneous Nature of the Process 151
   3.1 Grafting not Controlled by Diffusion 151
   3.2 Efficiency of Initiation and of Grafting 152
   3.3 Chain Termination in Heterogeneous Media 153
   3.4 Influence of the Grafted Polymer on the Process of Grafting 154

4 Special Features of Heterogeneous Graft Copolymerization 157

5 Basic Trends in Grafting to Cellulose Fibres in Industry 157

6 Polymer-Analogous Conversions of Functional Groups in Polymer Chains Grafted to Cellulose 162

7 References 164

List of Symbols

[c]   effective concentration of cellulose proportional to its mass
D   coefficient of monomer diffusion into the film
l   film width
k_g constant of chain growth
k_{ct} constant of chain termination
k_t constant of transfer to monomer
k_{rf} rate constant of the formation of cellulose radicals

k'_o rate constant of the decomposition of cellulose radicals
M   monomer concentration in the film
$\overline{P}$ average degree of polymerization
$\overline{R}$ averge grafting rate over the entire width of the film
R_i initiations rate
t   reaction time

1 Introduction

With decreasing oil resources and rising prices of oil and its products, interest in the production of polymer materials from cellulose, a highly important polymer, grows steadily. The vast supplies of cellulose could be still increased by rational arboriculture and utilization of agricultural wastes. The unlimited reproduction of cellulose is of economic benefit in that it consumes primarily solar energy and of ecological benefit, considering that plants absorb carbon dioxide from the atmosphere, purify air of solid particles, prevent soils from wind erosion, etc.

Therefore, the attention to the development of new polymer products based on cellulose or cellulose-containing raw materials will undoubtedly grow and the scope of their manufacture will be expanding.

From this standpoint, cellulose modification by grafting is exceptionally promising. Recently, grafting has become a popular method of modifying various high molecular weight compounds[1,2].

This process gives rise to products possessing the properties of both polymer components. Cellulose, in particular, may retain its valuable features such as high hydrophilicity, low electrifiability and considerable thermal stability and acquire new properties inherent to synthetic polymers.

Since the 1960's many researchers have been concerned with the development of feasible and industrially useful methods for the synthesis of cellulose graft copolymers[3,4]. Recent investigations have shown that the most efficient approach to this problem involves free radical polymerization initiated by redox systems[5]. An impressive example is the industrial production of mtilon (cellulose-polyacrylonitrile graft copolymer) and other fibers, particularly those with ion-exchange and acid-resistant properties[6-8].

Free-radical graft polymerization may be carried out in several ways[1,5]. However, only a few methods can be applied to the permanent production of graft-modified cellulose materials[9-12], and the development of industrial methods to ensure the rapid (30–60 s) grafting of the required amount of a polymer onto cellulose is still a major problem.

This review deals with recent results on the structure and properties of cellulose graft copolymers depending on the structure of cellulose, its pretreatment and the conditions of grafting. Also discussed are polymer-analogous conversions involving grafted chains.

2 Factors Influencing Graft Copolymerization with Cellulose, Structure and Properties of its Products

2.1 Effect of Cellulose Structure and Surface Area on the Kinetics of Grafting and Structure of the Products of Grafting

Grafting of any monomer to cellulose is a heterogeneous process essentially affected by the contact area between cellulose and monomer, the structure of cellulose and the rate of diffusion of monomer and initiator into the polymer. Structural and chemical changes in the primary polymer during grafting may change these parameters and their relationships, thus exerting a pronounced influence on the kinetics of the process.

The early stages of graft polymerization appear to be particularly affected by the structure and surface area of the initial cellulose material. In subsequent stages, the nature of the polymer being grafted and its swelling capacity in the reaction medium or in its own monomer are more important.

This observation may be illustrated by the results on grafting of styrene (St) onto cotton and viscose staple fiber. Figure 1 indicates that cotton, with its far greater specific surface area[13], sorbs much more styrene and the initial grafting rate is higher in this case. However, upon grafting of 10–12% polystyrene, the viscose fibre capacity for styrene sorption shows a drastic increase and the yield of graft copolymer grows, thus becoming comparable with that in the reaction with cotton. Thus, variation in the chemical composition of the fibre affects the monomer sorption, i.e. the concentration of styrene in the reaction zone. Monomers well soluble in water exhibit a different grafting behavior with respect ot cotton (itaconic acid[14] and acrylonitrile[15]). The specific surface area of the fibre is of little significance in such reactions since, apart from being sorbed, the monomer dissolved in water can diffuse into the cellulose fibre. The less ordered the structure is or, in other words, the lower the degree of crystallinity, the more accessible are the structural elements. For this reason, the use of viscose staple fibres results in substantially higher grafting rates, higher yields, of the grafted polymer and higher grafting efficiency than that of cotton.

This conclusion was confirmed by other experiments. Indeed, an increasing active surface area produced by grinding promotes grafting to cotton; however, the yield of the grafted polymer and the grafting efficiency were considerably lower than in the treatment with 40–80% ethylene diamine solution or 20% aqueous alkali[16, 17].

Studies on the hydrolysis of graft-modified cellulose have shown that the contant of the water-soluble fraction in this polymer increases with swelling during activation, apparently due to the decrease in the number of hydrogen bonds. Grafting to non-activated

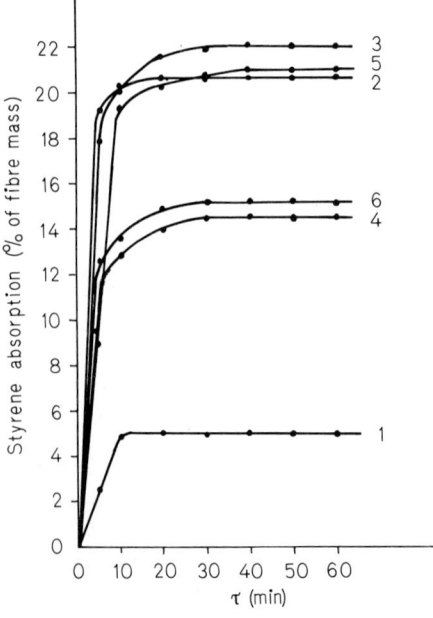

Fig. 1. Kinetic curves of styrene absorption. Curve 1: viscose staple fibre, curve 2: cotton, curve 3: graft copolymer of viscose staple fibre and polystyrene, composition 86.1% : 13.9%, curve 4: graft copolymer of cotton and polystyrene, composition 84.8% : 15.2%, curve 5: grafted polystyrene isolated from the copolymer with staple fibre, curve 6: grafted polystyrene isolated from the copolymer with cotton; absorption conditions: [St]–1%, M–25, emulsifier concentration 0.2%, temperature – 65°

ground cellulose is largely a surface process resulting in surface screening of cellulose structural elements by the grafted polymer and the corresponding deceleration of hydrolysis. The same authors have found[16] that 70–75% of poly(methyl methacrylate) is grafted to just 15% of ground cellulose.

Data on the effect of cellulose grinding on the yield of grafted polymers using acrylonitrile (AN) and ethyl acrylate (EA) reveal discrepancies[18, 19, 20]. While according to[18], the yield of grafted polymers increases with rising degree of grinding (i.e. rising specific surface area), according to[19], the yield is reported to decrease with grinding. On the other hand, the results obtained by Joseph[20] confirm the first statement. We presume that this discrepancy is due to the fact that different monomers were used so that the relationship between the grafting rate and the monomer diffusion rate was different in these studies.

2.2 Influence of the Cellulose Macroradical Accessibility on the Kinetics and Products of Graft Polymerization

The use of Ce^{4+}-based redox systems promotes cellulose oxidation and strongly accelerates grafting. Accordingly, at the very beginning of the process, diffusion into the fibre is hindered by a layer of grafted polymer appearing on the cellulose surface. Depending on the relationship between the grafting and the diffusion rates, therefore, grafting may occur at different "levels". For example, grinding of pre-irradiated cellulose increases the grafted polymer yield by making more accessible the radicals in the bulk of the fibre, especially in the crystalline regions. The relationship between the rates of grafting and diffusion thus governs both the grafted polymer distribution in the bulk of the fibre and the yield of grafted polymer. It was reported that the amount of PMMA grafted to merzerised cotton, non-merzerised cotton and cotton lint is lower than that grafted to viscose fibres (staple fibre, textile and cord thread) under the same conditions[21]. The authors suggest that in natural fibres, whose crystallinity is higher, the cellulose macroradicals resulting from irradiation are situated primarily in the crystalline regions where no diffusion of the monomer takes place.

The rate of decomposition of free cellulose macroradicals depends on the moisture content of the cellulose. Arthur et al.[22] observed that pre-irradiation of cotton produces a sufficient number of macroradicals if the primary polymer contains no more than 2% water. As shown by ESR, the number of radiation-induced macroradicals in cotton containing 6% water is three times as low as in cotton with 2% water.

In a Soviet study[23, 24] the mobility of cellulose macromolecular fragments was investigated by means of the paramagnetic label technique. Cellulose macroradicals serving as paramagnetic centres were obtained by irradiation of cellulose at $-120°$ to $-140°C$. The mobility of their fragments was found to increase sharply with the water content in the sample to reach a maximum at 10% water. It is this increase which appears to be responsible for the rapid decomposition of macroradicals in moist cellulose.

Arthur et al. reported that even in the presence of solvents in which cotton swells (HCL, dimethylformamide (DMFA) or dimethyl sulfoxide (DMSO)) about 30% of long-lived cellulose macroradicals resulting from irradiation were still unaccessible to the monomer. The same result was obtained from experiments[25] involving grafting of methacrylate (MA) and methyl methacrylate (MMA) to cellulose in the presence of water vapour upon irradiation in vacuum. It was also found that the amount of the grafted polymer decreased in the cotton-linenramie fibre series.

Initiation of graft copolymerization by the Fe^{2+}-H_2O_2 system in aqueous solution increases both the rate of AN grafting to cotton and the yield of grafted polymer. Initiation most probably involves lignin contained in the fibre. The lignin guajacic acid structure fragments rapidly reduce Fe^{3+} to Fe^{2+}, and this process accelerates the decomposition of hydrogen peroxide and increases the yield of ligno-cellulose complex macroradicals. On removal of lignin, the capacity of linen for grafting drops considerably.

Conversely, linen fibre pectins undergo graft polymerization[26] and their removal increases both the grafting rate and the yield of the grafted polymer. An additional reason for this effect is that pectin-free linen has a higher capillarity and a more accessible structure[27].

Lignins and pectins account for a considerable part of linen fibre. At the same time their oxidation occurs much more rapidly than that of cellulose. Accordingly, initiation by Ce^{4+} [27] and $KMnO_4$[28] led to lower reaction rates, and the yield of PAN grafted to linen was lower than in the case of the chain transfer method[27].

2.3 Grafting on Low-Substituted Cellulose Derivatives

The activation of cellulose by introducing new functional groups (even in a small number) which prevent the formation of hydrogen bonds and loosen the polymer structure may give rise to graft polymerization. For instance, alkali cellulose reacts with acryl amide to yield a 2-carbamoylethylated fibre containing 1 to 30% bound acryl amide[29]. Further grafting of alkyl acrylate to this material makes it completely amorphous and more elastic.

Low-substituted cellulose carbamate[30] and partially carboxymethylated or cyanoethylated cotton[31, 32] also proved highly accessible to grafting with in monomer the presence of an initiator. Grafting onto low-substituted acetate- and allyl cellulose is, however, less efficient because these materials only swell poorly in aqueous solutions or emulsions of the monomer[32, 33].

Ce^{4+}-initiated graft polymerization with viscose fibre and carboxymethylcellulose (CMC) was studied[34]. The molecular mass distribution (MMD) of grafted PMMA chains in both cases was described by a unimodal curve; with CMC the maximum was shifted towards higher molecular masses because of faster diffusion of the monomer into this material.

2.4 Cellulose Conversion During Graft Polymerization

The accessibility of cellulose to the monomer and the initiator may be estimated as the amount of cellulose incorporated into the graft copolymer[35]. Systematic investigations on the mutual influence of cellulose structure and the structure of the polymer being grafted were undertaken by Lifshits and Rogovin[36]. They found that PAN grafted to cellulose in aqueous solution involves about twice as much cellulose per unit weight of the grafted polymer as vapour-phase grafting. The viscose fibre studied is highly swellable in aqueous AN and up to 80% of the initial cellulose took part in grafting. The number of grafted chains per 1000 cellulose monomer units was 3–4 (for the amount of grafted PAN equal to 200% of the cellulose mass). The conversion of cellulose depends on the nature and the amount of the polymer to be grafted and on swelling (i.e. the accessibility of structural elements). Another crucial factor is the ratio between the rate of grafting and the rate of monomer diffusion into the ordered structures of the fibre. Oxidation-initiated grafting of AN to freshly formed (strongly swollen) viscose fibre with a residual content of thiocarbonyl groups y from 5 to 8[5]) resulted in a lower conversion than grafting to predried cellulose[35]) (Fig. 2). Figure 2 also indicates that the use of Fe^{3+} as a grafting initiator results in higher conversion of cellulose than V^{5+}. The average grafting rates were equal to 2% of the cellulose mass per minute with the H_2O_2-Fe^{2+} system used for initiation, up to 40% with the XC-Fe^{3+} system and up to 70% with the Xc-V^{5+} system. These results indicate that even in freshly formed swollen cellulose whose accessibility is far greater than that of the dry material, the conversion of cellulose decreases with increasing rate of grafting. This is evidently due to the fact that at high reaction rates the ordered supramolecular structure of cellulose is screened by grafted PAN chains, and the diffusion of monomer and initiator into the fibre is hindered.

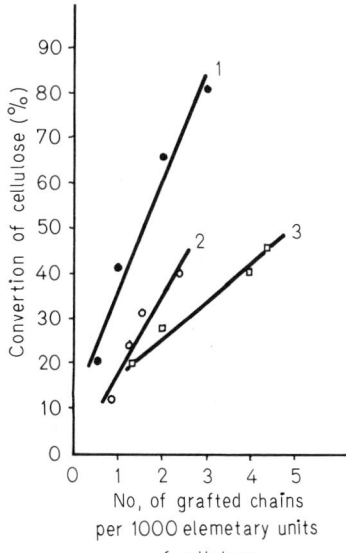

Fig. 2. Effect of the number of grafted chains on conversion of cellulose. Grafting initiated by: Curve 1: Fe^{2+} – H_2O_2 system; average grafting rate V_{av} – 2%/min, curve 2: CC – Fe^{3+} system; V_{av} – 40%/min, curve 3: CC – V^{5+} system; V_{av} – 70%/min

In an extensive X-ray diffraction study of cotton grafted with polyacrylamide (PAA) and poly(2-methyl-5-vinyl)pyridine (PMVP), Kurlyankina et al. have shown[37] that the size of crystallites in the initial cotton cellulose (55 Å) does not change upon grafting of either polymer at an amount of up to 900% of the cellulose mass. PAA grafting was found to increase the crystallinity of cellulose, as reported earlier[38]. This result, however, is contrary to that obtained by Lifshits and Rogovin who investigated the structure of graft copolymers of viscose staple fibre with PAN and PMMA and observed, in addition to a disordering of the fibre structure, a decrease in crystallinity (in the crystalline regions of the fibre) upon grafting. This discrepancy appears to stem from the smaller size and lower packing density of the staple fibre crytallites as compared to cotton. Thus, the diffusion of the monomer and the initiator into the staple fibre supramolecular molecules occurs more rapidly. Besides that, grafting monomers used in[37] (acrylamide (AA) and 2-methyl-5-vinylpyridine) increase swelling of the reaction product; the resulting higher mobility of cellulose macromolecular fragments leads to an increase in crystallinitiy. Conversely, grafting AN and MMA used by Lifshits and Rogovin decrease swelling and therefore promote formation of amorphous regions.

2.5 Influence of the Cellulose Matrix on the Orientation of the Grafted Chains

If neither cellulose nor the grafted polymer are capable of swelling in the reaction medium, the grafted chains acquire a certain orientation under the influence of the cellulose fibre structure[39]. The most remarkable example of this effect is furnished by vapour-phase radiation-induced grafting. Lifshits and Rogovin have observed the orientation effect even with swollen cellulose when the grafted polymer (PAN or poly(vinylidene chloride) (PVDCl)) did not swell in the reaction mixture[40]. Since grafting begins in the amorphous regions of the fibre, the orientation of grafted chains is not pronounced at the early stages of the process. Since the amount of the grafted polymer increases to 40–50% PAN, the degree of orientation rises; above this threshold, it starts to decrease again, due to disordering of the matrix. It is important to note that PAN chains are more strongly oriented when grafting conditions promote diffusion of monomer and initiator into the ordered regions of the fibre. For instance, PAN grafted in aqueous solutions of AN proves more oriented than that grafted in the vapour phase.

When diffusion into the ordered regions of the fibre during graft copolymerization is difficult, the grafted polymer structure may depend on submicroscopic pores in cellulose. Matsuzaki and Kanai[41] have reported that MMA grafting to pre-irradiated cellulose from methanol solution gave rise to a somewhat stereoregular product (with higher iso- and lower syndiotacticity) only with viscose threads, in contrast to wood cellulose and poly(vinyl alcohol). Similar results were obtained in a study on MMA grafting to cotton, cotton lint, copper ammonium silk, viscose silk and viscose staple fibre[25] under largely the same conditions as in[41].

2.6 Grafting on Cross-Linked Cellulose

Grafting to cross-linked cotton has been studied by several authors[42, 43]. Cross-linking makes the amorphous regions of the fibre (where grafting primarily takes place) less accessible to grafting and thus decreases both the grafting rate and the yield of the grafted polymer. An X-ray diffraction study of graft copolymers formed by certain acrylic monomers and cotton cross-linked by formaldehyde has shown[43] that grafting does not appreciably change the cellulose crystalline structure. The diffuse scattering of cross-linked cellulose after grafting increases somewhat more strongly than that of the non-treated material. The orientation of structural details after grafting decreases in both cases.

The effect of cross-linking on cotton graft polymerization with MA, vinyl acetate (VA), EA and MMA has been investigated quantitatively in[42]. Cellulose cross-linking was estimated from the content of bound formaldehyde which ranged from 0.022 to 0.42%; initiation was accomplished by Ce^{4+} salts. The cellulose containing 0.022% formaldehyde was more active in grafting than the initial cotton. In fact, the number of bridges generated in this process was insignificant while the accessibility of cellulose to grafting was higher due to phosphorous acid treatment during formylation.

The molecular mass of grafted chains decreases as the formaldehyde content in the fibre grows to 0.42% and then shows a slight increase which, as suggested in[42], is associated with the inhibition of chain termination involving cellulose hydroxy groups, whose accessibility to grafting decreases upon cross-linking. This explanation, however, is at variance with observations[44, 45] that in Ce^{4+}-initiated graft polymerization chain termination involves primarily oxidant cations. The extremum dependence reported in[42] may be explained as follows. In moderately cross-linked cellulose, there is a sufficiently large number of hydroxy groups to react with Ce^{4+} and give rise to the formation of macroradicals. Since these groups, however, are not readily accessible to grafting with monomer, a comparatively great number of short chains is formed. Further cross-linking decreases the number of OH groups in cellulose as well as the number of regions accessible to monomer diffusion. Under these conditions, a small number of relatively long chains is produced.

The effect of cotton cross-linking by formaldehyde and dimethylolurea on the yield of grafted methylol acrylamide and conversion of the monomer was studied in[46]. Graft copolymerization was photoinitiated. The cross-linking of unswollen cotton was found to decrease both parametres studied. In contrast, grafting to cotton cross-linked in the swollen state accounts for almost 100% conversion of the monomer. The reason is that cross-linking screens the loose structure of swollen cellulose thus making its structural elements more accessible to grafting with monomer.

Formation of transverse bonds between cellulose macromolecules during graft polymerization may change the grafting rate and the yield of the grafted polymer, depending on the length of the bridges. It was found, for instance, that Ce^{4+}-initiated grafting of AN and MMA to methylcarbamoyl cellulose involves cross-linking catalyzed by the acidic reaction medium[47]. Cross-linking decreases the accessibility of cellulose to grafting. Accordingly, the grafted polymer yield diminishes and the amount of homopolymer grows.

An opposite effect was observed in our studies on the synthesis of cellulose graft copolymers from viscose staple fibre and a binary mixture of acrylonitrile and 1,4-divinylbenzene (AN/DVB). The DVB cross-linking of grafted chains does not prevent the monomer from diffusion to cellulose active centres. At the same time, it decreases dramatically the mobility of grafted macroradicals, thus increasing their lifetime and providing for extremely rapid grafting and high conversion of the monomer.

It was demonstrated[48] that preliminary modification of cotton fabrics by grafting with AN, butyl acrylate (BA) or lauryl methacrylate enhances the efficiency of cross-linking by dimethyloldihydroxyethylurea, i.e. increases the tear resistance, abrasion resistance and the wrinkle recovery angle of the cross-linked fabric. Fabrics based on cross-linked copolymers of cotton with poly(methacrylic acid) proved less susceptible to contamination and showed easier oil and soil release than the cloths obtained from non-modified cross-linked cotton.

2.7 Morphology of the Products of Cellulose Grafting

Changing the relationship between the rate of diffusion and that of initiation is a powerful means of controlling the supramolecular structure of cellulose copolymers as well as the morphology of the modified fibre, i.e. the distribution of grafted chains in the bulk of the material.

The latter effect has not yet been investigated quantitatively. So far, only one interesting paper[49] by Odian and Kruse has been reported who performed a mathematical analysis of the grafting of styrene to irradiated polyethylene films for different ratios between the diffusion and the grafting rates. For stationary graft polymerization they derived the following equation:

$$(\overline{R}/MR_i^{1/2})(K_{ct/kg}^{1/2}) = \frac{\tank[(K_g/K_{ct}^{1/2})(R_i^{1/2}/D)]^{1/2} 1/2}{[(K/K_{ct}^{1/2})(R_i/q07^{1/2}/D)]^{1/2} 1/2} (\overline{R}\,l^2) \cdot 4\,MD =$$

$$[(K_g/K_{ct}^{1/2})(R_i^{1/2}/D)]^{1/2}\,1/2\,\tank[(K_g/K_{ct}^{1/2})(R_i^{1/2}/D)]^{1/2}\,1/2$$

This equation contains a dimensionless parameter

$$\alpha = [(K_g/K_t^{1/2})(R_i^{1/2}/D)]^{1/2}\,1/2$$

It was shown experimentally that if α is less than 0.1, graft polymerization occurs over the entire bulk of the film, and if α is greater than 3, only the film surface is involved in graft polymerization.

With chemical initiation, it is not easy to use this equation because graft polymerization results in screening of the functional groups, responsible for initiation, by the grafted polymer. Therefore, the increase in the amount of the grafted polymer changes the above-indicated parameters. Moreover, regardless of the initiation technique, the chemical composition of the fibre changes during grafting and so does the rate of monomer diffusion into the fibre (see, for instance, Fig. 1). These factors are difficult to be taken into account and hence the empirical approach to most problems concerning the distribution of graftet polymer in the cellulose material. However, the theoretical calculations reported by Odian and Kruse are undoubtedly of interest and may be employed for the directed search for graft polymerization conditions.

It has been shown in our laboratory that surface grafting to cellulose fibres is promoted by
a) higher grafting rates;
b) use of monomers poorly soluble in water and thus showing little diffusion into water-swollen cellulose;
c) swelling of the grafted polymer in its own monomer;
d) agents which cause swelling or plasticizing of the grafted polymer;
e) increase in the molecular mass of the grafted chains.

Figure 3 shows photomicrographs of viscose fibres containing grafted PAN (50% of the cellulose mass) after cellulose was etched away in such a manner that the chemical composition and the structure of the grafted polymer were not affected[50].

Grafting to these fibres was carried out under the following conditions:
a) from AN aqueous solution containing the initiating system H_2O_2–Fe^{2+};

Structural and Chemical Modifications of Cellulose by Graft Copolymerization

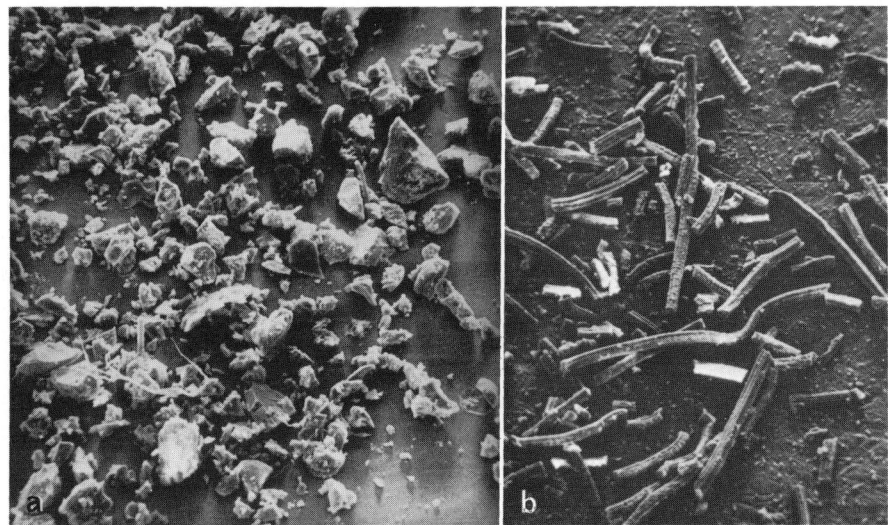

Fig. 3a, b. PAN-modified viscose fibres after etching. (**a**) 16% PAN, X 100, (**b**) 50% PAN, X 500

b) the same as in (a) but the fibre was freshly formed and contained a minor amount of thiocarbonyl groups ($\gamma \approx 0.5$);

c) the viscose threads were swollen, the cellulose incompletely regenerated so that it contained a small number of thiocarbonyl groups ($\gamma \approx 5$–7). Initiation was accomplished by oxidizing the SH groups with Fe^{3+} [12].

Under the conditions (a) the grafted PAN was uniformly distributed over the bulk of the fibre. In the case of freshly formed fibre, the grafting rate grows sharply from 2% of cellulose mass per minute to 10%, due to the participation of active SH groups in chain transfer[51]; the diffusion rate grows but only slightly since both types of fibres exhibit approximately the same degree of swelling in water. The increase in the grafting rate soon results in a grafted polymer "shell" which covers the fibre surface and prevents diffusion of the monomer and the initiator into the bulk.

The rate of grafting to strongly swollen incompletely regenerated viscose fibre is lower than the rate of diffusion (an assumption indirectly confirmed by the fact that the diffusion rate of acids into such fibre is by one order of magnitude higher than the rate of hydrolysis of its thiocarbonyl groups[52]), and the grafted PAN again shows uniform distribution over the bulk of the fibre. On the other hand, grafting of MVP, MMA and St from aqueous emulsions is primarily a surface process since these monomers are incapable of rapidly diffusing into water-swollen cellulose.

Data on the effect of the amount of the grafted polymer (PAN) on its distribution in the cellulose fibre are given in Fig. 3. Upon removal of cellulose, the samples containing 15–20% PAN disintegrate into individual microblocks, apparently due to the fact that the bonds between these species in the initial sample were weak. This observation indicates that at early stages of grafting (15–20 min) there is no continuous reaction front directed from the surface into the bulk of the fibre and that the grafted polymer forms individual blocks whose size and number increase during the reaction. The generation of the new blocks occurs as the formation of new chain growth centres ceases due to hindered diffusion of the initiator (and the monomer in the case of PAN) into the regions rich in grafted polymer. The free initiator and the monomer continue to penetrate into more ordered areas not involved in the early stages of grafting and after accumulating there in nesufficient amounts they give rise to initiation and growth of grafted chains or, in other words, to the formation of new blocks. Therefore, the increase in the amount of the grafted polymer promotes uniform distribution of grafted chains. The most convincing illustration is the grafting of polymers which do not swell in the reaction medium and in their own monomer (e.g. PAN). The uniform distribution in this case is self-controlled owing to slower diffusion of both the initiator and the monomer into the fibre regions already involved in grafting.

Radiation-initiated grafting to swollen cellulose results in a grafted polymer located largely in the bulk of the fibre. Surface grafting under these conditions is prevented by the reaction between atmospheric oxygen and the surface cellulose macroradicals which thus decompose rapidly and are not involved in initiation. A comparative morphological study of hydrated cellulose fibres modified by radiation ethyl acrylate grafting and by chemical grafting of the same monomer was reported in[53]. Irradiation gave rise to a grafted polymer uniformly distributed over the bulk of the fibre; the grafted chains produced by chemical initiation were located primarily in the surface layer. The authors observed that upon decrystallization, the fibres obtained by radiation grafting and containing about 100% PEA manifested pronounced hyperelastic properties, while no such effect was established in the case of chemically initiated grafting. This fact may be explained as follows: in the former case, the grafted chains essentially affect the initial fibre structure due to their uniform distribution between cellulose macromolecules (grafting is closer to the molecular level).

2.8 Influence of the Grafted Polymer Distribution on the Properties of the Grafted Copolymers

Arthur et al.[54] have shown that cotton modified by AN grafting retains the structure and appearance of the initial fibre (particularly the transverse cross-section shape) much better if the grafted polymer is located in the surface layer. To this end, they suggest graft polymerization to cotton pre-irradiated in an inert atmosphere and the use of solvents causing no swelling of the fibre.

For grafting to involve the entire bulk of the cotton fibre the matrix should swell in the reaction medium. This can be achieved by the use of appropriate solvents or agents decreasing interactions between cellulose macromolecules; another means is the use of low-substituted cellulose derivatives more susceptible to swelling. For instance, AN bulk grafting to cotton included cyanoethylation of the fibre, treatment with AN solution in a solvent causing strong swelling and, finally, irradiation.

Similar results were reported for radiation grafting (Co^{60}-rays) of AN and St to viscose fibres[55]. The fibres swollen in the methanol solution of the monomer were irradiated in air and the grafted polymer was formed essentially in the bulk. Surface grafting is necessary when the cellulose fibre is to acquire ion-exchange properties[7] or become resistant to aggressive media[56].

The authors of this review have found that a fibre covered by grafted and then alkylated poly(2-methyl-4-vinylpiridine) ensures extremely rapid ion exchange virtually independent off diffusion. In contrast, ion exchange with the fibre formed from the graft copolymer solution[57] and containing grafted chains rather uniformly distributed over the bulk proved appreciably sensitive to diffusion factors.

A continuous defect-free "shell" formed by the grafted polymer on the surface of viscose staple fibre makes the fibre remarkably resistant to aggressive media[58]. Such a material was obtained by grafting 15–20% PS to the fibre and subsequent treatment with trichloroethylene which caused swelling of the grafted PS and thus healing of the surface layer defects. It is interesting to note that absorption of water vapour by this fibre (9–10%) is just slightly weaker than that by the regular viscose staple fibre (11–12%); at the same time, the fibre is not wetted by water and by acid solutions. It appears that water vapour diffuses into cellulose through microscopic pores in grafted PS blocks (see photograph) which, however, are too narrow to allow penetration of liquids with their high surface tension.

Stannet et al.[59] have shown that the grafting of P- and Br-containing vinyl monomers to cellulose yields products with the highest fire resistance if the grafted chains are uniformly distributed over the bulk of the fibre. To achieve this the initial fibre was allowed to swell at 30 °C for 30 min and then kept in an antipyrene solution for 2 h. After excessive monomer was removed from the surface by decantation, the fibre was irradiated. Rapid antipyrene treatment immediately followed by irradiation resulted in a product containing grafted chains primarily on the surface.

The uniform distribution of the grafted polymer in the bulk of the fibre and its deep penetration into the ordered regions apparently give rise to significant changes in the physico-mechanical properties of modified cellulose. The penetration of the grafted polymer into microcracks of the fibre and into intercrystalline areas of the structure causes appreciable stresses which result, for example, in a wool-like appearance and low wettability of the viscose staple fibre graft copolymer with rigid-chain PAN. Surface grafting does not lead to such effects.

To sum up, by changing the distribution of the grafted polymer in the cellulose fibre one can control the properties of the modified material.

3 Kinetics of Chemically Initiated Graft Copolymerization and its Features Arising from the Heterogeneous Nature of the Process

3.1 Grafting not Controlled by Diffusion

Graft polymerization with cellulose is always a heterogeneous process markedly dependent on diffusion factors. In most cases, however, especially when a minor amount of the polymer is grafted, these factors are neglected and the process is regarded as taking place kinetically controlled. For the graft polymerization of several vinyl monomers (AA, MVP, acrylic acid (AAC) and methacrylic acid (MAAc)) initiated by the cellulose-Co^{3+} redox system, such kinetic equations not taking into account the diffusion factors were derived by Kurlyankina et al.[60–62].

The two initiation methods most efficient for grafting of monomer and polymer to cellulose involve oxidation of the matrix by variable valency metal ions and chain transfer from the initiator radicals to cellulose. The oxidant cations or the initiator radicals in these cases usually take part in chain termination[44, 62]. Proceeding from this fact, the following equation has been suggested[62].

$$\frac{1}{\overline{P}} = \frac{k_{ct} [Co(III)]}{k_g [M]} + \frac{k_t}{k_g}$$

Apart from reacting with the growing grafted chain macroradicals, the oxidant cations or the initiator radicals may cause decomposition of cellulose macroradicals formed at the first stage of oxidation. Taking into account the latter process, the rate of graft polymerization initiation may be described as follows[62]:

$$\frac{1}{V_i} = \frac{1}{k_{rf} [C] [Co^{3+}]} + \frac{k'_o}{k_t [M] k_{rf} [C]}$$

k_g/k_{ct} = 0.24 × 10² for acrylamide (AA) and 0.28 for methylvinylpyridine (MVP);
k_t/k_g = 4 × 10⁻⁴ for AA and 1.5 × 10⁻³ for MVP
k_t/k_{ct} = 3 × 10⁻³ for AA and 1.6 × 10⁻³ for MVP

3.2 Efficiency of Initiation and of Grafting

The initiation efficiency depends on the relationship between the rate of macroradical formation and that of initiation. In Table 1 are compiled results on certain redox systems employed for the initiation of cellulose grafting either by direct oxidation of cellulose (or its derivative) or chain transfer from active low molecular weight radicals. Table 1 indicates that in systems where the matrix acts as a reducing agent (1st group), the initiation efficiency does not exceed 15%, i.e. only a minor portion of macroradicals formed at the first stage of oxidation initiates graft polymerization while the rest is oxidized to stable

Table 1. Certain redox systems used in cellulose graft polymerization

System	Redox potential of the oxidant E_0, (V)	Grafting efficiency (%)	Initiation efficiency (%)	Max. grafting rate, % of cellulose (mass/min.)	Ref.
Co^{3+}-cellulose	up to 1.8	70–80	8–10	–	62)
Ce^{4+}-cellulose	1.55	80–90	8–10	10	44)
Mn^{3+}-cellulose	1.4	60–70	10–15	6	63)
V^{5+}–cellulose–C₆H₄–NH₂ 	1.0	90–95	8–9	–	64)
V^{5+}–cellulose–OC(S)SH	1.0	90–95	10–11	60	65)
Fe^{3+}–cellulose–O–C(S)SH	0.76	95–98	3–4	40	12)
H_2O_2-Fe^{2+}-cellulose		94–97	25–30	3	5)
H_2O_2-Fe^{2+}-hydroquinone-cellulose		95–96	25–30	70	66)
H_2O_2-Fe^{2+}-hydrazine-cellulose		95–96	35–50	20	66)

products by the metal ions. Such low efficiency was attributed to the fact[62] that the rate constant of the monomer (e.g. MVP or AA) addition to cellulose macroradicals is 400–600 times as low as the rate constant of the macroradical interaction with the metal ion. Initiation is much more efficient if it is based on chain transfer from active OH radicals (H_2O_2-Fe^{2+} systems) to cellulose. In such systems, the stationary concentration of OH radicals is low and, accordingly, the [M]/OH ratio is higher than the corresponding value [M]/[Me^{n+1}] in the direct oxidation methods. It also seems likely that chain transfer requires a substantially lower activation energy than direct cellulose oxidation. Therefore, the rate of OH radical interaction with cellulose cannot differ greatly from the rate of the interaction of OH radicals with cellulose macroradicals. For example, the activation energy of Co^{3+}-initiated grafting of MVP is equal to 90.4 kJ/mol[62] which is fairly close to the activation energy of cellulose oxidation by Co^{3+}. The total activation energy of graft polymerization initiated by the H_2O_2-Fe^{2+} system is as low as 14 kJ/mol.

As shown in Ref. 5, some cellulose macroradicals react with Fe^{3+} ions to decrease the initiation efficiency. This undesirable effect is suppressed in reversible redox systems where the Fe^{3+} ions are rapidly reduced by an additional component. As a result, the probability of the decomposition of cellulose macroradicals decreases and the initiation efficiency grows to 50%; moreover, the generation of OH radicals is accelerated which permits grafting at a lower temperature.

The grafting efficiency is determined as the ratio of the amount of the grafted polymer to the sum of this value and the amount of the homopolymer. This practically important parameter depends on the rates of the reactions of Me^{n+1} cations (or OH radicals) with cellulose and with the monomer. The Me^{n+1} ions whose oxidation potential is low (Fe^{3+}, V^{5+}) virtually do not react with the monomer and therefore cause no homopolymerization. On the contrary, ions with rather high oxidation potentials such as Co^{3+}, Ce^{4+} and Mn^{3+} do initiate homopolymerization after a certain induction period. Systems in which the polymer is a sufficiently strong reducing agent and the potential of the oxidant is low are evidently to be preferred in the design of feasible grafting methods.

Chain transfer initiation may be promoted by immobilizing one of the components (usually the reducing agent) on cellulose in order to diminish the probability of reaction between the low molecular weight radical and the monomer. The best results were obtained with systems generating a highly active, small amount of OH radical[63].

3.3 Chain Termination in Heterogeneous Media

Chain termination during graft polymerization does not always involve interactions between growing macroradicals and oxidant ions or initiator radicals. Other mechanisms may be observed with hydrophobic polymers and with polymers not swelling in the reaction medium, as well as at late stages of grafting when there is no initiator available or the oxidant left in the system. Lifshits reported[67] that cellulose graft copolymer with a high PAN content, which was obtained under the conditions of complete exhaustion of the H_2O_2 supply in the system, contains a cross-linked fraction formed by cellulose chains connected by PAN macromolecules. This effect is due to chain termination by recombination of grafted PAN macroradicals. It was also observed[68] that emulsion grafting of styrene results in grafted chains of high molecular mass (up to 1.5 mln) which is almost independent of H_2O_2 concentration.

As long as grafting takes place in such viscous environment as cellulose, the rate of chain termination is much lower than in the case of homopolymerization. Grafting is facilitated, however, if the polymer macromolecule contains groups highly susceptible to chain transfer reactions. A good example is grafting to wool keratin rich in mercapto groups[69]. With cellulose derivatives, this mechanism was reported for grafting using sodium periodate[70] when chain termination involved formyl groups of the matrix. The low molecular masses of the resulting grafted polymers suggest the same mechanism of chain termination for grafting using redox systems which contain xanthogenate cellulose[65, 71].

VA grafting at 45 °C may be completely suppressed by ascorbic acid[72]. It is well-known that the H_2O_2-Fe^{2+}-ascorbic acid system rapidly generates free radicals even at low temperatures and may be used for efficient grafting at room temperature[73]. At 45 °C the initiation rate was obviously so high that the reaction time of the monomer was short and virtually all the cellulose macroradicals reacted with the initiator radicals.

At high initiation rates the rate of grafting shows an appreciable increase if the temperature is raised and thus the monomer diffusion into the fibre promoted[68]. It is interesting to note that the acceleration of grafting (e.g. in the case of AN) is due both to the formation of longer chains and to the increase in the number of chains. The former effect, which contradicts the laws of regular radical polymerization, indicates that grafting of the polymers highly capable of swelling in the reaction medium is essentially dependent on the diffusion of the monomer.

The grafting rate dependence on the rate of monomer diffusion into the grafted fibre in non-aqueous media was reported for photo-initiated grafting of styrene to cellulose in the presence of various alcohols[74]. The highest reaction rate and the highest yield of grafted polymer were observed in the 1:1 methanol-isobutanol mixture which causes swelling of both the cellulose and the grafted polystyrene. In methanol, which does increase the reaction rate as compared to that in higher alcohols, the effect was less pronounced than in the previous medium because, in contrast to cellulose, the grafted PS does not swell in methanol.

3.4 Influence of the Grafted Polymer on the Process of Grafting

Graft polymerization decreasing the reactivity of the cellulose material (by decreasing its swelling, thus forming a continuous grafted polymer shell on its surface or by hydrophobization) is a diffusion-sensitive process even at early stages. In the case of AN grafting from aqueous solutions initiated by the H_2O_2-Fe^{2+} system, the H_2O_2 decomposition rate and, accordingly, the rate of initiation decrease greatly as the content of grafted PAN reaches 20–30% (Fig. 4). The reason for this effect which, in the case of PS, is observed if the polymer content amounts to 10–15%, is the screening of Fe^{2+} ions by the grafted chains. If the grafted polymer absorbs the monomer from the solution or emulsion, the reaction is controlled by the diffusion of the initiator (e.g. H_2O_2) into the fibre. If the grafted polymer hinders diffusion of the initiator and the monomer alike (AN grafting), the reaction rate depends primarily on the rate of the monomer diffusion. Indeed, at high concentrations of the initiator, the grafting initiation efficiency decreases considerably owing to the decrease in the number of graftet chains. Thus, the grafted

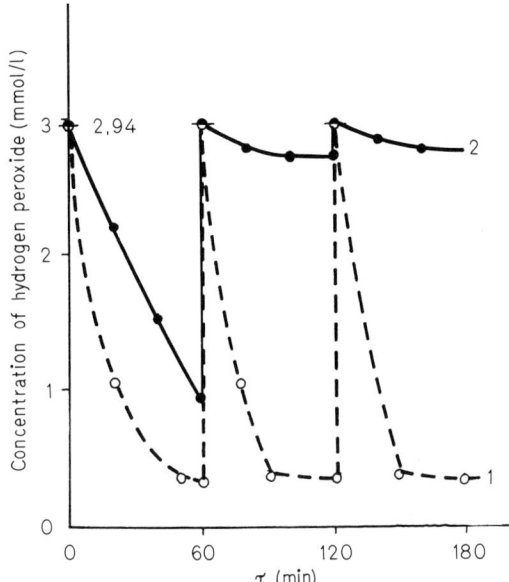

Fig. 4. Kinetic curves of hydrogen peroxide decomposition under the conditions used for grafting. Curve 1 – grafting takes place, curve 2 – no monomer is present. H_2O_2 – 2.94×10^{-3} M/l, [AN] – 1.4 M/l, M – 50, temperature – 70 °C

polymer yield vs. the initiator concentration curve has a maximum. If the reaction zone inside the fibre lacks monomer, most cellulose macroradicals would decompose by reaction with initiator radicals. Poorly soluble or chemically inert monomers used for grafting impressively illustrate this effect. As shown by the authors, the acceleration of initiation without raising the temperature (e.g. by adding a reducing agent such as hydrazine or hydroquinone to the H_2O_2-Fe^{2+} system) may considerably decelerate styrene grafting or even stop it altogether.

The influence of more than 100 organic media (solvents for the monomer) on radiation grafting of 2,4-dimethyl- and 2-methyl-5-vinylpyridine to cellulose (filter paper of various brands and degrees of grinding) was shown to be associated with two factors, namely cellulose swelling and specific interaction of the organic liquid with the monomer and with cellulose[75, 76].

The acidic and basic properties of the organic reagent are crucial to both the monomer diffusion and to grafting initiation.

The above data suggest that the grafting rate may be controlled by diffusion of either the monomer or the initiator into the fibre. The former effect occurs at the first stage of polymerization, when the amount of the grafted polymer is small, the latter at the final stages.

Grafting of rigid chain polymers auch as PAN which do not swell in the reaction medium at low concentrations of the initiator may give rise to long-lived grafted macroradicals (occluded macroradicals).

It was found that cellulose graft copolymers containing 30–50% PAN can, after the reaction medium has been washed out with water, initiate graft polymerization of various vinyl monomers (Table 2) without any additional initiators. No effect was observed with test samples of cellulose treated under the same conditions as those of PAN grafting but without AN.

Table 2. Polymerization of vinyl monomers initiated by occluded macroradicals of PAN grafted to cellulose

No.	PAN content in the graft copolymer	Conditions of the initiation				Amount of second monomer grafted	Copolymer composition	
		monomer	M (%)	τ (min)	t (°C)		PAN (wt-%)	M (wt-%)
1	69.0	2-methyl-5-vinylpyridine	100	180	85	40.0	42.0	58.0
2	34.0	2-methyl-5-vinylpiridine	15	180	85	2.0	94.1	5.9
3	48.0	Ca acrylate	25.0	120	70	20.0	58.0	42.0
4	40.2	methacrylic acid	7.0	120	70	10.2	76.8	23.2
5	48.0	methacrylic acid	100	120	70	20.0	58.4	41.6
6	34.0	methyl methacrylate	100	120	70	8.0	76.5	23.5

PAN grafting conditions (1st stage): Fe^{2+} = 0.73 mg/g fibre, [AN] = 7%, [H_2O_2] = 0.005%, t = 70°C, τ = 30 min

4 Special Features of Heterogeneous Graft Copolymerization

The foregoing aspects of graft copolymerization with cellulose as a heterogeneous process are crucial to the use of this method in technology, e.g. for the synthesis of graft-modified hydrated cellulose fibres[7, 68].

Grafting of certain polymers well absorbing their own monomer (e.g. St or MVP) changes the chemical composition of the substrate in such a manner that a fibre containing 10–20% grafted polymer (of the cellulose mass) starts to absorb the monomer. The grafted polymer chains swell and form a continuous surface layer which decelerates diffusion of the water-soluble initiator (e.g. H_2O_2) into the fibre. As a result, the reaction slows down and the monomer conversion decreases. Further addition of initiator does not improve the situation. This difficulty, however, may be overcome if, upon grafting of 15–30% polymer, an organic hydroperoxide capable of easy diffusion through the grafted layer (for instance, isopropylbenzene hydroperoxide) is added to the system[77]. Such an approach, which increases the monomer conversion to 90–99.9%, was used for the industrial production of CM-A 2 ion-exchange fibre and modified viscose staple fibre containing 15–20% grafted polystyrene and possessing high resistance ot aggressive media as well as excellent hygienic properties[68].

The diffusion of monomer into the grafted fibre makes it difficult to control the polymerization process by the consumption of monomer and promotes non-uniform grafting. In view of these circumstances, grafting technology should involve preliminary uniform treatment of the fibre by the monomer emulsion and provide for relatively slow graft polymerization at the early stages of the process.

The distribution of the grafted chains over the fibrous mass as a whole and over unit fibres is essentially more uniform if the grafted polymer (e.g. PAN or PVDCl) does not absorb its own monomer from the reaction medium. In this case, the rate of diffusion of monomer and initiator into non-grafted cellulose material exceeds that into the areas containing grafted chains. Unfortunately, the method of increasing the monomer conversion just described for the grafting of polymers absorbing their own monomer cannot be applied here because at late stages of grafting, the monomer diffusion into the fibre becomes very slow.

We have shown that with polymers not absorbing their own monomer it is possible to utilize the formation of long-lived grafted occluded macroradicals. These are efficiently formed if as many chains as possible are grafted at the beginning of the process. The starting concentration of the initiator (e.g. H_2O_2), the temperature and the mode of treatment must be selected in such a manner that within 30 min from the beginning of the reaction, the initiator would decompose completely, while the solution would still contain enough monomer (about 30% of the starting monomer). Upon decomposition of the initiator, the temperature should be raised to ensure access of monomer to occluded macroradicals. Graft polymerization then continues at sufficiently a high rate up to 80–85% conversion of monomer. This technique was successfully used for the industrial production of mtilon fibre.

5 Basic Trends in Grafting to Cellulose Fibres in Industry

Experience on the industrial production and use of mtilon and, of late, other fibres based on cellulose graft copolymers (e.g. ion-exchange fibres and fibres resistant to aggressive media) has shown that their applications are highly economic. The growing demand for these fibres requires a continuous process of their production. A first step in this direction was the creation of a semi-continuous process for the synthesis of mtilon on the basis of the existing periodic one in the USSR. Advances in the development of rapid methods of grafting polymers to viscose fibres have essentially made it possible to devise methods of continuous production of graft copolymers possessing various composition and wider fields of application. Viscose fibres may be simultaneously obtained and modified only provided the grafting rate is not lower than 10–30% polymer (of the cellulose mass) per minute and no homopolymerization occurs. These requirements are met by the methods based on the use of xanthogenate cellulose and such oxidants as V^{5+}, Fe^{3+} and Cr^{6+} [5] as well as by the processes recently developed in our laboratory and involving initiation by reversible redox systems[66].

Figure 5 describes a flow sheet of grafting to freshly formed viscose fibre containing a minor amount of thiocarbonyl groups (the degree of XC substitution is 0.1–0.15). Incom-

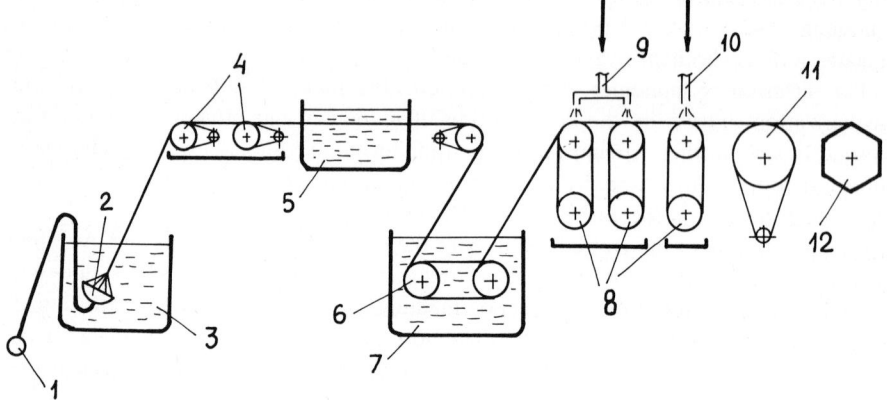

Fig. 5. Schematic representation of a spinning mill for the manufacture of modified viscose fibre. 1: viscose collector, 2: spinneret assembly, 3: precipitation bath, 4: spinning discs, 5: plastisizing bath, 6: loading roller, 7: grafting bath, 8: finishing rollers, 9: rinsig water feed, 10: brightening solution feed, 11: drying cylinder, 12: receiving assembly

Table 3. Effect of reaction time on the amount of PAN, PMAAc and PAAc grafted during the shaping process

Reaction time (s)	Grafted monomer		
	AN	MAAc	AAc
	Grafted PAN, % of cellulose mass	Grafted PMAA, % of cellulose mass	Grafted PAAc, % of cellulose mass
17	8.5	–	–
30	19.0	31.8	14.6
60	–	47.7	18.8
120	52.0	57.0	25.0
200	75.0	61.0	32.0

Grafting efficiency 95–98%

pletely regenerated fibre is fed to a grafting bath containing the monomer and the oxidant (V. Fe, Cr). On leaving the bath, the fibre passes a compartment where excessive monomer is removed, and then undergoes a routine finishing cycle. Table 3 contains data characterizing the process for the production of different graft-modified viscose fibres. Significantly, this process can yields fibres of various types (thread, cord or fibrous mass).

Reversible redox system ensure grafting to viscose fibres at any stage of technological treatment of the shaped material. Such systems have been used as sources of free radicals to initiate homopolymerization of various monomers[78, 79]. A detailed study has been reported by Dolgoplosk et al.[80]. The systems consist of a peroxide initiator (usually an organic hydroperoxide), a metal of variable valency catalyzing the peroxide decomposition and an agent capable of rapidly reducing the metal ion.

In theory, such systems may be used for highly efficient grafting, due to very rapid generation of active initiator radicals and the absence of the inhibiting effect of the metal ions reacting with cellulose macroradicals[5]. The actual use of reversible systems involves a number of difficulties, due to the high probability of homopolymerization on account of the interaction between the peroxide and the third component (reducing agent). It was suggested to avoid homopolymerization by immobilizing the third component on the cellulose material[81]. Unfortunately, this method reduces the quantity of the reducing agent in the system and therefore the grafting rate[82]. At the same time, recent investigations have shown that certain reducing agents cause no homopolymerization even during prolonged contact with monomer. These agents include hydroquinone, hydrazine, sodium sulfide, sodium pyrophosphate, glucose, and sodium bitartrate.

Data in Table 4 describe the efficiency of the three-component systems in the synthesis of cellulose – PAN graft copolymers. The reaction rate is highest when hydroquinone is used as an additional reducing agent. Hydroquinone rapidly reduces Fe^{3+} ions even at low temperatures[83, 84] and permits grafting at lower temperatures than the other agents; even at $-17\,°C$ is the graft polymerization of AN fast enough (Table 5).

The acceleration of grafting caused by hydroquinone increases the number of grafted chains but makes them shorter. The reaction rate and the yield of the graft polymer increase, however, since the former effect dominates the latter (Table 6). The chain shortening is due to the higher stationary concentration of OH radicals involved in chain termination and to the fact that hydroquinone and its oxidation products (semiquinone and benzoquinone) may also cause decomposition macroradicals.

Table 4. Effect of additional reducing agents on the amount of grafted PAN

Reducing agent	Concentration		Amount of grafted PAN, % of initial cellulose mass
	of agent, $M \times 10^{-3}$	of H_2O_2, $M \times 10^{-3}$	
–	–	2.94	29
Hydroquinone	1.82	2.94	105
Hydrazine	0.312	2.94	94
Na sulfide	0.41	14.7	60
Glucose	0.114	14.7	38
Na hypophosphite	0.114	14.7	41
Na bitartrate	0.114	14.7	50

Reaction conditions: [AN]: 7%, t: 60 °C, τ: 20 min, [Fe^{2+}]: 0,25%, M = 50

Table 5. Effect of hydroquinone on the yield of grafted PAN under various conditions

		Initiating system	
		$Fe^{2+}-H_2O_2$	$Fe^{2+}-H_2O_2$-hydroquinone
Reaction time (min)	Temperature (°C)	Amount of grafted PAN, (%) of cellulose mass	Amount of grafted PAN, % of cellulose mass
330	−17	−	43
60	20	−	91
60	40	5.0	104
1[x)]	60	−	46
1[x)]	70	−	60
1[x)]	77	−	70

Reaction conditions: [AN]: 7%, [AN[x)]]: 9%, [H_2O_2]: 0.01%, [Fe^{2+}]: 0.25%, [hydroquinone]: 0.02%, M = 50

Table 6. Effect of H_2O_2 and hydroquinone concentration on molecular mass of grafted PAN chains and their number

Time (min)	Concentration C × 10⁻³ (mol/l)		Amount of grafted PAN, % of initial cellulose mass	Molecular mass	n, number of chains per 1000 units cellulose	a, number of moles of decomposed H_2O_2 per 1000 cellulose units, (mol/l)	f[a] initiation efficiency (%)
	H_2O_2	hydroquinone					
40	2.94	−	46	97,720	0.76	6.25	24.4
20	2.94	0.182	67	56,230	1.97	6.4	61.6
20	2.94	0.364	94.1	50,120	3.05	10.4	58.8
30	2.94	1.82	106	25,700	6.7	23.8	56.4
20	2.94	4.54	62.2	23,990	4.2	23.8	35.2
30	14.7	1.82	111	23,990	7.5	50.2	29.8
30	29.4	1.82	128	25,800	8.1	65.6	24.6
60	147	1.82	117	19,100	10	252	7.92
60	441	1.82	31.9	4,000	12.9	453	5.7

[a] The initiation efficiency was estimated by means of $f = \dfrac{n \times 100}{a \times 1/2}$. Reaction conditions: [AN] = 7%: M = 50: [Fe^{2+}] = 0.25%

We have found that hydroquinone and its oxidation products are primarily responsible for chain termination. Table 7 indicates that at equal stationary concentrations of OH radicals, the molecular mass of grafted chains depends on the concentration of hydroquinone rather than that of hydrogen peroxide.

A cellulose graft copolymer containing 55% PAN was shown to contain 5.4 grafted chains per 1000 cellulose units, every other chain terminating in a quinone fragment as a result of the interaction between a macroradical and a semiquinone radical.

Semiquinone also decreases the initiation efficiency by reaction with cellulose macroradicals; grafting of every PAN chain is accompanied by the decomposition of about 5 cellulose macroradicals.

Table 7. Effect of hydroquinone concentration on the molecular mass of grafted PAN chains

Hydroquinone concentration (%)	H_2O_2 decomposition rate (mol $l^{-1}min^{-1} \times 10^4$)	Amount of grafted PAN (%)	Molecular mass of grafted chains	Number of grafted chains per 1000 cellulose units
0.02	2.2	45.8	42,660	1.74
0.05	2.0	44	11,220	6.35

Reaction conditions: t = 60 °C: [AN]: 7%

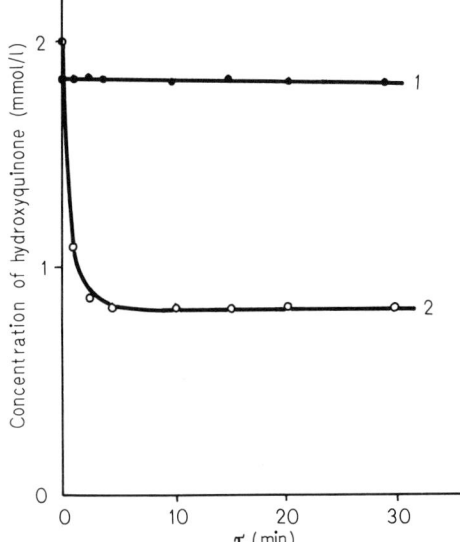

Fig. 6. Kinetic curves of hydroquinone oxidation: 1: H_2O_2, 2: Fe^{3+}. Reaction conditions: $[Fe^{3+}]$ – 0.25%, M – 50, temperature – 60 °C

Hydroquinone and benzoquinone apparently participate in chain termination and react with cellulose macroradicals. It seems likely, however, that the main role in these processes is played by semiquinone. Indeed, upon penetration into the fibre, hydroquinone reacts first with Fe^{3+} (the reaction with H_2O_2, as Fig. 6 indicates, is extremely slow under the conditions studied) to yield semiquinone. This reaction proceeds much more rapidly than subsequent oxidation of semiquinone to benzoquinone by Fe^{3+}[83], and one may expect that the stationary concentration of semiquinone in the fibre exceeds those of hydroquinone and benzoquinone. The inhibiting role of semiquinone in the polymerization of non-saturated compounds was pointed out by Dolgoplosk[85].

The addition of semiquinone to cellulose macroradicals gives rise to low-substituted derivatives containing 5 to 10 hydroquinone groups per 100 cellulose units.

It is interesting to note that the system containing hydroquinone does not initiate homopolymerization of aqueous AN under the conditions identical to those of grafting. Dolgoplosk attributed this effect to inhibition by hydroquinone oxidation products and suggested that inhibition can be suppressed by a fourth component reducing these products. The initiation of graft polymerization by this system appears to arise from the fact that chain termination in water-swollen cellulose is difficult.

The use of reversible redox systems as initiators allows to intensify the periodic grafting process[66, 86] and to develop a continuous technology for the synthesis of modified fibres. For instance, when using such systems, the rate of AN grafting to viscose staple fibre (Table 5) may be as high as 70% PAN of the cellulose mass per minute.

6 Polymer-Analogous Conversions of Functional Groups in Polymer Chains Grafted to Cellulose

Polymer-analogous conversions of polymer chains grafted to cellulose have been widely studied. They involve introduction of ionogenic groups[87-92], formaldehyde cross-linking of polyacrylamide chains grafted to cotton[93] or PAN chains after alkaline hydrolysis[94], and the synthesis of cross-linked fibres from cellulose-polyacrolein graft copolymers[95].

Kamogawa and Sekija[93] carried out Ce^{4+}-initiated grafting of acrylamide to cotton, methylolated the amide groups with 10% aqueous formaldehyde at pH 11 and cross-linked the product with ethylene urea, melamine, adipamide, glycerol, and certain other agents. The copolymers cross-linked with ethylene urea and melamine possessed the highest break and tear resistance Gordon[96] grafted N-methylolacrylamide to cellulose and found that the methylol groups of the grafted chains react with cellulose in the presence of an acidic catalyst.

Kulkarni et al.[94] subjected cotton-PAN graft copolymer to alkaline hydrolysis, methylolated the resulting product at pH 9.5–10 with formaldehyde for 24 h and cross-linked the polymer in an acidic medium at 150 °C in the presence of $MgCl_2$ for 5 min. This treatment increased the wrinkling resistance of the cotton fabric appreciably. No effect, however, was observed if the grafted chains were cross-linked by 1,4-divinylbenzene simultaneously with grafting.

Japanese workers[92] studied partial alkaline hydrolysis of wood cellulose graft copolymers using poly(vinyl acetate) (PVA) to enhance the interactions between fibres in paper manufacture. The amount of the grafted PVA was 10–45% with respect to the cellulose mass, the extent of hydrolsis about 30%. The paper produced from such modified cellulose has a smooth surface and high strength in both the dry and wet state. In[97] alkaline hydrolysis of cellulose-AN copolymers containing 40–100% PAN was carried out with 1–4% NaOH solution for 5 min; the product is subsequently kept in a steam bath. It was found that the ratio between the amounts of carboxy and amide groups in grafted chains after hydrolysis depends on that between the concentration of NaOH and the PAN content. The ion exchange capacity of the product reached 2.7 mg eq./g.

Pavlov et al.[98] observed that partially hydrolyzed cellulose-PAN copolymers lend themselves to dyeing by acid and direct dyes. Unfortunately, the products fail to form fibres suitable for textile industry because of their poor mechanical properties.

A Soviet study[99] was concerned with grafting of 1-vinyl-2-pyrrolidone glycidyl methacrylate copolymer to cotton. The product underwent polymer-analogous conversions through interaction of its reactive epoxy groups with diethylamine, thiourea, isoitaconic acid hydrazide and sodium hydrogen sulfite. The maximum conversion of the epoxy groups was 97.8, 90.2, 95.8 and 48.9% respectively. The resulting preparations may be of interest as physiologically active compounds.

Graft copolymers of cellulose with glycidyl methacrylate yield cation-exchange materials with sodium hydrogen sulfite[100] and cationites with ammonia, monoethanolamine and diethylamine.

The epoxy groups of poly(glycidyl methacrylate) grafted to cellulose were used for the chemical addition of dyes containing amino groups or phenolic hydroxy groups[101].

As a rule, fibrous ionites obtained by the conversion of functional groups of grafted chains were not strong enough for textile processing due to considerable swelling during the polymer-analogous reaction. To overcome this difficulty it was recommended, for instance, to hydrolyze grafted PAN and certain acrylic polymers with concentrated alkali solutions at 100–130 °C for 10–30 min[102]. The resulting fibre, however, was still too swellable.

Textile processing of modified fibres may become possible by improving their physico-mechanical properties by 1,4-divinylbenzene cross-linking of grafted ionogenic polymers or of polymers whose functional groups are readily converted to ionogenic ones. For instance, alkaline hydrolysis of PAN grafted to viscose staple fibre in the presence of DVB yields a product which contains 20% carboxy groups and can be easily processed to a non-woven material. A similar effect was achieved by grafting with methylvinylpyridine in the presence of DVB and subsequent N-alkylation with dimethyl sulfate. Without cross-linking, the product was highly swollen and could not be used for textile processing.

Interesting results were obtained in a comparative study on the conversion of the cyano groups of PAN chains grafted to cellulose and the conversion of the cyano groups of highly substituted cellulose cyanoethyl ester to amidoxyme and aminothiocarbonyl groups[7, 88]. Both the reaction rates and the yields were appreciably higher in the case of graft copolymers. In samples containing more than 40% PAN (with respect to cellulose mass), the yield of aminothiocarbonyl groups decreased. Recent investigations have shown that the increase in the AN-DVB content of grafted chains also decreases the yield of carboxy groups resulting from alkaline hydrolysis.

The effect reported in [7, 88] was attributed to the higher hydrophilicity of the graft copolymers as compared to the cellulose esters. The authors did not allow for the role played by the grafted polymer structure in these reactions. At the same time, it could be interesting to investigate the conversion of cyano groups in the graft copolymer and in PAN chains isolated from the copolymer so that their supramolecular and molecular structures would remain intact.

Such investigations performed by the authors concerned nitration of cellulose-PS graft copolymer, PS isolated from this product and PS obtained by suspension homopolymerization of styrene. The grafted PS was isolated so as to preserve its molecular mass, chemical composition and supramolecular structure[104]; the homopolymer was reprecipitated from benzene into methanol. The molecular mass of grafted chains was 8×10^5–10^6 and that of the homopolystyrene 5×10^5.

Figure 7 indicates that the homopolymer nitration rate was very low. In contrast, nitration of grafted chains under the same conditions is a rapid process somewhat decelerating with increasing PS content of the samples. The PS isolated from the copolymer is nitrated much more slowly. It seems therefore that the high nitration rate of PS grafted chains is due to topochemical factors, i.e. the specific distribution of the chains in the copolymer, rather than to the larger contact area between PS and the nitrating agent.

It is also interesting to note that the nitration of cellulose in the graft copolymer with PS occurs much more slowly than that of the cotton cellulose used for grafting (Fig. 7).

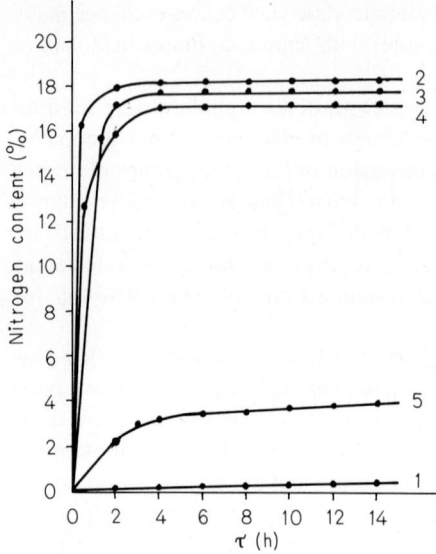

Fig. 7. Kinetic curves of nitration. 1: PS homopolymer, 2: PS grafted to cellulose (74.6% : 25.4%), 3: same as 2, composition (60.6% : 39.4%), 4: same as 2, composition (52.1% : 47,9%), 5: PS chains isolated from the graft copolymer containing 39.4% PS. Reaction conditions: nitrating mixture HNO_3 : CH_3COOH and $(CH_3CO)_2O$ in the ratio 50 : 25 : 25 (mass %), temperature − 10°C, M − 25

The cellulose nitration rate decreases with rising content of PAN in the sample.

These data point to the specific behaviour of grafted PS chains in polymer-analogous conversions. As a rule, the functional groups of grafted macromolecules appear to be far more reactive than their homopolymer counterparts.

The elucidation of the mechanism which accounts for the nitration behaviour of cellulose-PS graft copolymers, particularly the mutual influence of the copolymer components in polymer-analogous conversions, is a problem of great scientific interest.

7 References

1. Rogovin, Z. A., Gal'braykh, L. S.: Chemical Conversions and Modifications of Cellulose, Moscow, Khimiya, 1979, p. 205 (in Russian)
2. Zeltobrjuchow, V. F. et al.: Faserforsch. Textiltechn. *30*, 236 (1979)
3. Livshits, R. M., Rogovin, Z. A.: Uspekhi Khimii *34*, 1086 (1965)
4. Livshits, R. M., Rogovin, z. A., in: Advances in Polymer Chemistry, Moscow, Nauka, 1969, p. 158 (in Russian)
5. Morin, B. P., Rogovin, Z. A.: Vysokomol. Soed. *18 A*, 2147 (1976)
6. Cherepanov, V. N. et al.: Khim. Volokna *5*, 68 (1973)
7. Lishevskaya, M. O., Morin, B. P., Rogovin, Z. A.: Abstracts of Papers Presented at the 1st (Internat. Symp. on Chemical Fibres, Kalinin, 1974, p. 122 (in Russian)
8. Bereza, M. P., Morin, B. P., Rogovin Z. A.: Tekstil'naya Promyshlennost' *1*, 74 (1973)
9. US Pat. 3 445 556 (1967)
10. Samoilov, V. I. et al.: Khim. Volokna *3*, 33 (1974)
11. Samoilov, V. i. et al.: Khim. Volokna *1*, 41 (1976)
12. Morin, B. P., Rogovin, Z. A., Samoilov, V. I.: Vysokomol. Soed. *19 B*, 772 (1976)
13. Mel'nikov, B. N., Blincheva, I. B.: Theoretical Introduction to Technology of Fibrous Materials Dyeing, Moscow, Legkaya Industria, 1978, p. 303

14. Dimov, K. et al.: Cell. Chem. and Technol., 9, 6, 575 (1975)
15. Morin, B. P., Livshits, R. M., Rogovin, Z. A.: Vysokomol. Soed. *10A*, 875 (1967)
16. Kurosu, H., Horiike, K.: J. Jap. Wood Res. Soc. *22*, 444 (1976)
17. Kurosu, H., Horiike, K., Mokuzai, G.: J. Jap. Wood Res. Soc. *22*, 358 (1976)
18. Monsour, o., Nagaty, A.: J. Appl. Polym. Sci. *23*, 2445 (1979)
19. Ogiwara, Y., Kubotu, H.: Kogyo Kagaku Zasshi *71*, 1, 171 (1968); C. A. *C8*, 115 817 (1968)
20. Erderlyi, J.: Papiri Par. *15*, 10 (1971); C. A. *75*, 8 94 08 (1971)
21. Matsuzaki, K., Kanai, T.: Cell. Chem. Technol. *12*, 413 (1978)
22. Arthur, J. C. Jr., Harris, J. A., Mares, T.: Text. Industr. *132*, 77 (1968)
23. Buchachenko, A. L., in: Stable Radicals, Moscow, AN SSSR, 1963, p. 172
24. Plotnikov, O. V., Mikhailov, A. I., Rajavse, E. L.: Vysokomol. Soed. *11*, 2528 (1977)
25. Azizov U., Mirkamilov, I., Usmanov, Kh.: Radiation Graft Polymerisation, Tashkent, Rap. 1976 (in Russian)
26. Ekman, K., Enkwist, T.: Pap. ja puu, *37*, 369 (1955)
27. Shekhterman, E. I. et al.: Zh. Prikl. Khim. *42*, 1425 (1969)
28. El-Shinnary, N., Allan, E., Hebeish, A.: Cell. Chem. Technol. *13*, 565 (1979)
29. Nakamura, Y., Miamoto, K.: Jap. Pat. 49-42 368 (1976)
30. Hebeish, A. et al.: Angew. Makromol. Chem. *70*, 87 (1978)
31. Hebeish, A. et al.: J. Appl. Polym. Sci. *17*, 2547 (1973)
32. Hebeish, A., Kantouch, A., El-Rafie, M. H.: ibid. *15*, 11 (1971)
33. Hebeish, A., El-Rafie, A., El-Aref, A. T.: Angew. Makromol. Chem. *78*, 195 (1979)
34. Kaputsky, F. N., Siderko, V. M., Starostina, O. I.: Izv. Akad. Nauk. Beloruss. SSR 121 (1975)
35. Sydykov, T. S. et al.: Khim. Volokna *4*, 76 (1967)
36. Sydykov, T. S. et al.: Vysokomol. Soed. *8*, 2035 (1966)
37. Volkova, l. A. et al.: Vysokomol. Soed. *B20*, 892 (1978)
38. Simonescu, Gr., Oprea, S.: Cell. Chem. Technol. *4*, 71 (1970)
39. Tseytlin, B. A., in: Radiation Chemistry of Polymers, Moscow, Nauka, 1973, p. 151
40. Rogovin, Z. A. et al.: Vysokomol. Soed. *B10*, 461 (1968)
41. Matsuzaki, K., Kanai, T.: J. Appl. Polym. Sci. *20*, 2221 (1976)
42. Varma, D. S., Narasimhan, V.: J. Appl. Polym. Sci. *19*, 29 (1975)
43. Varma, D. S., Narasimhan, V.: Indian J. Text. Res. *1*, 60 (1976)
44. Rogovin, Z. A., Livshits, R. M.: Buletinul Institutului Politehnic Din Lasi, 1970, XVI (XX), No. 1–2, Sectia II, Chemie Industiala, P. 297
45. Kurlyankina, V. I., Sarana, N. V., Koz'mina, O. P.: Kinetika i Kataliz, *11*, 5 (1970)
46. Reinhardt, R. M., Arthur, J. C. Jr., Muller, L. L.: J. Appl. Polym. Sci. *24*, 1739 (1979)
47. Abou Zeid, N. Y., Anwar, W., Hebeish, A.: Cell. Chem. Technol. *14*, 203 (1980)
48. Harris, J. A. et al.: Textile Ind. *133*, 117 (1969)
49. Odian, G., Kruse, R.: Prepr. of the Internat. Symp. on Macromol. Chem. Brussel-Louvain, June 12–16, 1967, p. 142
50. Bereza, M. P., Morin, B. P., Rogovin, Z. A.: Khim. Volokna *1*, 42 (1977)
51. Channdhure, R. D. K., Hermans, G. G.: J. Polym. Sci. *49*, 159 (1960)
52. Zakharov, V. S., Zelentsov, I. G., Pahkver, A. B.: Khim. Volokna *3*, 28 (1960)
53. Williams, J. L. et al.: Int. J. Appl. Radiat. Isotop. *26*, 169 (1975)
54. Arthur, J. C. Jr.: Macromol. Sci. *A10*, 653 (1976)
55. Staroverova, L. L., Kabanov, V. Ya.: Khim. Volokna *3*, 20 (1979)
56. Bereza, M. P. et al.: Cell. Chem. Technol. *11*, 293 (1977)
57. Wilkov, V. I., Morin, B. P., Rogovin, Z. A.: Faserforsch. Textiltechn. *24*, 77 (1973)
58. Bereza, M. P. et al.: Faserforsch. Textiltechn. *26*, 564 (1975)
59. Stannett, Y. et al.: Appl. Polym. Symp. *1977*, 201
60. Kurlyankina, V. I., Molotkov, V. A., Koz'mina, O. P.: Vysokomol. Soed. *B11*, 117 (1972)
61. Molotkov, V. A., Kurlyankina, V.I., Klenin, S. I.: Vysokomol. Soed. *A14*, 2478 (1972)
62. Kurlyankina, V. I. et al.: Vysokomol. Soed. *A18*, 997 (1976)
63. Livshits, R. M., Morin, B. P., Rogovin, Z. A.: Cell. Chem. Technol. *1*, 153 (1967)
64. Livshits, R. M., Predvoditelev, D. A., Rogovin, Z. A., in: Cellulose and its Derivatives, AN SSSR, 1963, p. 60
65. Samoilov, W. I., Morin, B. P., Rogovin, Z. A.: Faserforsch. Textiltech. *22*, 297 (1971)

66. Voinova, G. Yu. et al.: Khim. Volokna *3*, 30 (1980)
67. Gulina, A. A., Livshits, R. M., Rogovin, Z. A.: Vysokomol. Soed. *7*, 1529 (1965)
68. Morin, B. P., Bereza, M. P., Rogowin, Z. A.: Papier *31*, 365 (1977)
69. Breusova, I. P. et al.: Vysokomol. Soed. *A 21*, 196 (1979)
70. Toda, T.: J. Polym. Sci. *58*, 414 (1962)
71. Dimov, K., Pavlov P.: J. Polym. Sci. *7*, 2775 (1960)
72. Misra, B. N. et al.: J. Polym. Sci., Polym. Chem. Ed. *17*, 1861 (1979)
73. Zhdanova, Yu. P. et al.: Cell. Chem. Technol. *10*, 315 (1976)
74. Davis, N. P., Garnett, J. L., Urguhart, R.: J. Polym. Sci., Polym. Lett. Ed. *14*, 537 (1976)
75. Garnett, J. L., Martin, E. C.: J. Polym. Sci., Polym. Lett. Ed. *14*, 35 (1976)
76. Garnett, J. L., Martin, E. C.: Austral. J. Chem.: *29*, 2991 (1976)
77. Morin, B. P. et al.: USSR Pat. 444 773 (1973) B. I. *36*, 1975
78. Tryling, C. F., Follett, A.: J. Polym. Sci. *6*, 59 (1951)
79. Kolthoff, i. M., Medalia, A. I.: J. Polym. Sci. *6*, 189 (1951)
80. Dolgoplosk, B. A., Tinyakova, E. I.: Redox Systems as Sources of Free Radicals, Moscow, Nauka, 1972, p. 238 (in Russian)
81. Morin, B. P. et al.: USSR Pat. 401 675 (1972), B.I. 41, 1974
82. Bershova, N. V. et al.: Promyshlennost Khimicheskikh Volokon, inf. abstr., 1975
83. Kassidi, G. D., Kun, K. A.: Redox Polymers, Leningrad, Khimiya, 1967, p. 269 (in Russian)
84. Baxendale, Y. H., Hardy, H. R., Lutchiffe, L. H.: Trans. Faraday Soc. *47*, 963 (1951)
85. Dolgoplosk, B. A., Tinyakova, E. I.: Khimich. Nauka i Prom. *2*, 280 (1957)
86. Voinova, G. Yu. et al.: Khim. Volokna *2*, 36 (1980)
87. Vladimirova, T. V. et al.: Izv. VUZov, Khim. i khimich. tekhnol. *XI*, 594 (1967)
88. Rogovin, Z. A. et al.: US Pat. 3 728 103 (1974) and 3 82 11 37 (1974)
89. Mazov, M. Yu., Tyuganova, M. A.: Cell. Chem. Technol. *10*, 185 (1976)
90. Akbarova, M. G., Ibragimov, R. I., Gafurov, T. G.: Dokl. AN Uzb. SSR *1977*, 33
91. Gal'braikh, L. S., Vladimirova, T. V., Rogovin, Z. A.: Izv. VUZov, Khim. i khimich. technol. *9*, 144 (1966)
92. Vashimi, M., Sakamoto, I., Ohiro, H.: Jap. appl. No. 53-74105 of Dec. 7, 1976, No. 51-177 355 publ. July 1, 1978
93. Kamogawa, H., Sekija, T.: Text. Res. J. *31*, 585 (1961)
94. Kulkarni, A. Y. et al.: J. Appl. Polym. Sci. *7*, 1581 (1963)
95. Maedzima, T., Yamashita, N., Asakura, D.: Jap. Pat. 5 341 274 (1974); 4 958 274 (1978)
96. Gordon, J. L.: J. Appl. Polym. Sci. *5*, 734 (1961)
97. Ehnhort, E. et al.: Pap. ja Puu *59*, 11 (1977)
98. Pavlov, P., Simeonov, N., Dimov, K.: Godishn. Vis. Khim.-Tekhn. Ins., Sophia *23*, 73 (1977)
99. Sharkova, E. F., Virnik, A. D., Rogovin, Z. A.: Vysokomol. Soed. *6*, 951 (1964)
100. Kirasaki, H., Iwakura, E.: Khimiya i Techn. Polim. *10*, 65 (1963)
101. Cornell, R. H.: Tappi *45*, 145 A (1962)
102. French Pat. 2 250 765 (1976)
103. US Pat. 4 198 326 (1980)
104. Bennet, C. F., Timmell, T. E.: Svensk. Papperstidn. *58*, 281 (1955)

Received May 4, 1981
G. V. Schulz (editor)

Author Index Volumes 1–42

Allegra, G. and *Bassi, I. W.:* Isomorphism in Synthetic Macromolecular Systems. Vol. 6, pp. 549–574.
Andrews, E. H.: Molecular Fracture in Polymers. Vol. 27, pp. 1–66.
Anufrieva, E. V. and *Gotlib, Yu. Ya.:* Investigation of Polymers in Solution by Polarized Luminescence. Vol. 40, pp.1–68.
Ayrey, G.: The Use of Isotopes in Polymer Analysis. Vol. 6, pp. 128–148.
Baldwin, R. L.: Sedimentation of High Polymers. Vol. 1, pp. 451–511.
Basedow, A, M. and *Ebert, K.:* Ultrasonic Degradation of Polymers in Solution. Vol. 22, pp. 83–148.
Batz, H.-G.: Polymeric Drugs. Vol. 23, pp. 25–53.
Bekturov, E. A. and *Bimendina, L. A.:* Interpolymer Complexes. Vol. 41, pp. 99–147.
Bergsma, F. and *Kruissink, Ch. A.:* Ion-Exchange Membranes. Vol 2, pp. 307–362.
Berlin, Al. Al., Volfson, S. A., and *Enikolopian, N. S.:* Kinetics of Polymerization Processes. Vol. 38, pp. 89–140.
Berry, G. C. and *Fox, T. G.:* The Viscosity of Polymers and Their Concentrated Solutions. Vol. 5, pp. 261–357.
Bevington, J. C.: Isotopic Methods in Polymer Chemistry. Vol. 2, pp. 1–17.
Bird, R. B., Warner, Jr., H. R., and *Evans, D. C.:* Kinetik Theory and Rheology of Dumbbell Suspensions with Brownian Motion. Vol. 8, pp. 1–90.
Biswas, M. and *Maity, C.:* Molecular Sieves as Polymerization Catalysts. Vol. 31, pp. 47–88.
Block, H.: The Nature and Application of Electrical Phenomena in Polymers. Vol. 33, pp. 93–167.
Böhm, L. L., Chmeliř, M., Löhr, G., Schmitt, B. J. und *Schulz, G. V.:* Zustände und Reaktionen des Carbanions bei der anionischen Polymerisation des Styrols. Vol. 9, pp. 1–45.
Bovey, F. A. and *Tiers, G. V. D.:* The High Resolution Nuclear Magnetic Resonance Spectroscopy of Polymers. Vol. 3, pp. 139–195.
Braun, J.-M. and *Guillet, J. E.:* Study of Polymers by Inverse Gas Chromatography. Vol. 21, pp. 107–145.
Breitenbach, J. W., Olaj, O. F. und *Sommer, F.:* Polymerisationsanregung durch Elektrolyse. Vol. 9, pp. 47–227.
Bresler, S. E. and *Kazbekov, E. N.:* Macroradical Reactivity Studied by Electron Spin Resonance. Vol. 3, pp. 688–711.
Bucknall, C. B.: Fracture and Failure of Multiphase Polymers and Polymer Composites. Vol. 27, pp. 121–148.
Bywater, S.: Polymerization Initiated by Lithium and Its Compounds. Vol. 4, pp. 66–110.
Bywater, S.: Preparation and Properties of Star-branched Polymers. Vol. 30, pp. 89–116.
Carrick, W. L.: The Mechanism of Olefin Polymerization by Ziegler-Natta Catalysts. Vol. 12, pp. 65–86.
Casale, A. and *Porter, R. S.:* Mechanical Synthesis of Block and Graft Copolymers. Vol. 17, pp. 1–71.
Cerf, R.: La dynamique des solutions de macromolecules dans un champ de vitesses. Vol. 1, pp. 382–450.
Cesca, S., Priola, A. and *Bruzzone, M.:* Synthesis and Modification of Polymers Containing a System of Conjugated Double Bonds. Vol. 32, pp. 1–67.
Cicchetti, O.: Mechanisms of Oxidative Photodegradation and of UV Stabilization of Polyolefins. Vol. 7, pp. 70–112.
Clark, D. T.: ESCA Applied to Polymers. Vol. 24, pp. 125–188.

Coleman, Jr., L. E. and *Meinhardt, N. A.:* Polymerization Reactions of Vinyl Ketones. Vol. 1, pp. 159–179.
Crescenzi, V.: Some Recent Studies of Polyelectrolyte Solutions. Vol. 5, pp. 358–386.
Davydov, B. E. and *Krentsel, B. A.:* Progress in the Chemistry of Polyconjugated Systems. Vol. 25, pp. 1–46.
Dole, M.: Calorimetric Studies of States and Transitions in Solid High Polymers. Vol. 2, pp. 221–274.
Dreyfuss, P. and *Dreyfuss, M. P.:* Polytetrahydrofuran. Vol. 4, pp. 528–590.
Dušek, K. and *Prins, W.:* Structure and Elasticity of Non-Crystalline Polymer Networks. Vol. 6, pp. 1–102.
Eastham, A. M.: Some Aspects of the Polymerization of Cyclic Ethers. Vol. 2, pp. 18–50.
Ehrlich, P. and *Mortimer, G. A.:* Fundamentals of the Free-Radical Polymerization of Ethylene. Vol. 7, pp. 386–448.
Eisenberg, A.: Ionic Forces in Polymers. Vol. 5, pp. 59–112.
Elias, H.-G., Bareiss, R. und *Watterson, J. G.:* Mittelwerte des Molekulargewichts und anderer Eigenschaften. Vol. 11, pp. 111–204.
Fischer, H.: Freie Radikale während der Polymerisation, nachgewiesen und identifiziert durch Elektronenspinresonanz. Vol. 5, pp. 463–530.
Fujita, H.: Diffusion in Polymer-Diluent Systems. Vol. 3, pp. 1–47.
Funke, W.: Über die Strukturaufklärung vernetzter Makromoleküle, insbesondere vernetzter Polyesterharze, mit chemischen Methoden. Vol. 4, pp. 157–235.
Gal'braikh, L. S. and *Rogovin, Z. A.:* Chemical Transformations of Cellulose. Vol. 14, pp. 87–130.
Gallot, B. R. M.: Preparation and Study of Block Copolymers with Ordered Structures, Vol. 29, pp. 85–156.
Gandini, A.: The Behaviour of Furan Derivatives in Polymerization Reactions. Vol. 25, pp. 47–96.
Gandini, A. and *Cheradame, H.:* Cationic Polymerization. Initiation with Alkenyl Monomers. Vol. 34/35, pp. 1–289.
Geckeler, K., Pillai, V. N. R., and *Mutter, M.:* Applications of Soluble Polymeric Supports. Vol. 39, pp. 65–94.
Gerrens, H.: Kinetik der Emulsionspolymerisation. Vol. 1, pp. 234–328.
Ghiggino, K. P., Roberts, A. J. and *Phillips, D.:* Time-Resolved Fluorescence Techniques in Polymer and Biopolymer Studies. Vol. 40, pp. 69–167.
Goethals, E. J.: The Formation of Cyclic Oligomers in the Cationic Polymerization of Heterocycles. Vol. 23, pp. 103–130.
Graessley, W. W.: The Etanglement Concept in Polymer Rheology. Vol. 16, pp. 1–179.
Hagihara, N., Sonogashira, K. and *Takahashi, S.:* Linear Polymers Containing Transition Metals in the Main Chain. Vol. 41, pp. 149–179.
Hasegawa, M.: Four-Center Photopolymerization in the Crystalline State. Vol. 42, pp. 1–49.
Hay, A. S.: Aromatic Polyethers. Vol. 4, pp. 496–527.
Hayakawa, R. and *Wada, Y.:* Piezoelectricity and Related Properties of Polymer Films. Vol. 11, pp. 1–55.
Heitz, W.: Polymeric Reagents. Polymer Design, Scope, and Limitations. Vol. 23, pp. 1–23.
Helfferich, F.: Ionenaustausch. Vol. 1, pp. 329–381.
Hendra, P. J.: Laser-Raman Spectra of Polymers. Vol. 6, pp. 151–169.
Henrici-Olivé, G. und *Olivé, S.:* Kettenübertragung bei der radikalischen Polymerisation. Vol. 2, pp. 496–577.
Henrici-Olivé, G. und *Olivé, S.:* Koordinative Polymerisation an löslichen Übergangsmetall-Katalysatoren. Vol. 6, pp. 421–472.
Henrici-Olivé, G. and *Olivé, S.:* Oligomerization of Ethylene with Soluble Transition-Metal Catalysts. Vol. 15, pp. 1–30.
Henrici-Olivé, G. and *Olivé, S.:* Molecular Interactions and Macroscopic Properties of Polyacrylonitrile and Model Substances. Vol. 32, pp. 123–152.
Hermans, Jr., J., Lohr, D. and *Ferro, D.:* Treatment of the Folding and Unfolding of Protein Molecules in Solution According to a Lattic Model. Vol. 9, pp. 229–283.
Holzmüller, W.: Molecular Mobility, Deformation and Relaxation Processes in Polymers. Vol. 26, pp. 1–62.

Hutchison, J. and *Ledwith, A.:* Photoinitiation of Vinyl Polymerization by Aromatic Carbonyl Compounds. Vol. 14, pp. 49–86.

Iizuka, E.: Properties of Liquid Crystals of Polypeptides: with Stress on the Electromagnetic Orientation. Vol. 20, pp. 79–107.

Ikada, Y.: Characterization of Graft Copolymers. Vol. 29, pp. 47–84.

Imanishi, Y.: Syntheses, Conformation, and Reactions of Cyclic Peptides. Vol. 20, pp. 1–77.

Inagaki, H.: Polymer Separation and Characterization by Thin-Layer Chromatography. Vol. 24, pp. 189–237.

Inoue, S.: Asymmetric Reactions of Synthetic Polypeptides. Vol. 21, pp. 77–106.

Ise, N.: Polymerizations under an Electric Field. Vol. 6, pp. 347–376.

Ise, N.: The Mean Activity Coefficient of Polyelectrolytes in Aqueous Solutions and Its Related Properties. Vol. 7, pp. 536–593.

Isihara, A.: Intramolecular Statistics of a Flexible Chain Molecule. Vol. 7, pp. 449–476.

Isihara, A.: Irreversible Processes in Solutions of Chain Polymers. Vol. 5, pp. 531–567.

Isihara, A. and *Guth, E.:* Theory of Dilute Macromolecular Solutions. Vol. 5, pp. 233–260.

Janeschitz-Kriegl, H.: Flow Birefringence of Elastico-Viscous Polymer Systems. Vol. 6, pp. 170–318.

Jenkins, R. and *Porter, R. S.:* Unpertubed Dimensions of Stereoregular Polymers. Vol. 36, pp. 1–20.

Jenngins, B. R.: Electro-Optic Methods for Characterizing Macromolecules in Dilute Solution. Vol. 22, pp. 61–81.

Johnston, D. S.: Macrozwitterion Polymerization. Vol. 42, pp. 51–106.

Kamachi, M.: Influence of Solvent on Free Radical Polymerization of Vinyl Compounds. Vol. 38, pp. 55–87.

Kawabata, S. and *Kawai, H.:* Strain Energy Density Functions of Rubber Vulcanizates from Biaxial Extension. Vol. 24, pp. 89–124.

Kennedy, J. P. and *Chou, T.:* Poly(isobutylene-*co*-β-Pinene): A New Sulfur Vulcanizable, Ozone Resistant Elastomer by Cationic Isomerization Copolymerization. Vol. 21, pp. 1–39.

Kennedy, J. P. and *Delvaux, J. M.:* Synthesis, Characterization and Morphology of Poly(butadiene-*g*-Styrene). Vol. 38, pp. 141–163.

Kennedy, J. P. and *Gillham, J. K.:* Cationic Polymerization of Olefins with Alkylaluminium Initiators. Vol. 10, pp. 1–33.

Kennedy, J. P. and *Johnston, J. E.:* The Cationic Isomerization Polymerization of 3-Methyl-1-butene and 4-Methyl-1-pentene. Vol. 19, pp. 57–95.

Kennedy, J. P. and *Langer, Jr., A. W.:* Recent Advances in Cationic Polymerization. Vol. 3, pp. 508–580.

Kennedy, J. P. and *Otsu, T.:* Polymerization with Isomerization of Monomer Preceding Propagation. Vol. 7, pp. 369–385.

Kennedy, J. P. and *Rengachary, S.:* Correlation Between Cationic Model and Polymerization Reactions of Olefins. Vol. 14, pp. 1–48.

Kennedy, J. P. and *Trivedi, P. D.:* Cationic Olefin Polymerization Using Alkyl Halide – Alkylaluminum Initiator Systems. I. Reactivity Studies. II. Molecular Weight Studies. Vol. 28, pp. 83–151.

Khoklov, A. R. and *Grosberg, A. Yu.:* Statistical Theory of Polymeric Lyotropic Liquid Crystals. Vol. 41, pp. 53–97.

Kissin, Yu. V.: Structures of Copolymers of High Olefins. Vol. 15, pp. 91–155.

Kitagawa, T. and *Miyazawa, T.:* Neutron Scattering and Normal Vibrations of Polymers. Vol. 9, pp. 335–414.

Kitamaru, R. and *Horii, F.:* NMR Approach to the Phase Structure of Linear Polyethylene. Vol. 26., pp. 139–180.

Knappe, W.: Wärmeleitung in Polymeren. Vol. 7, pp. 477–535.

Koningsveld, R.: Preparative and Analytical Aspects of Polymer Fractionation. Vol. 7.

Kovacs, A. J.: Transition vitreuse dans les polymers amorphes. Etude phénoménologique. Vol. 3, pp. 394–507.

Krässig, H. A.: Graft Co-Polymerization of Cellulose and Its Derivatives. Vol. 4, pp. 111–156.

Kraus, G.: Reinforcement of Elastomers by Carbon Black. Vol. 8, pp. 155–237.

Kreutz, W. and *Welte, W.:* A General Theory for the Evaluation of X-Ray Diagrams of Biomembranes and Other Lamellar Systems. Vol. 30, pp. 161–225.

Krimm, S.: Infrared Spectra of High Polymers. Vol. 2, pp. 51–72.

Kuhn, W., Ramel, A., Walters, D. H., Ebner, G. and *Kuhn, H. J.:* The Production of Mechanical Energy from Different Forms of Chemical Energy with Homogeneous and Cross-Striated High Polymer Systems. Vol. 1, pp. 540–592.

Kunitake, T. and *Okahata, Y.:* Catalytic Hydrolysis by Synthetic Polymers. Vol. 20, pp. 159–221.

Kurata, M. and *Stockmayer, W. H.:* Intrinsic Viscosities and Unperturbed Dimensions of Long Chain Molecules. Vol. 3, pp. 196–312.

Ledwith, A. and *Sherrington, D. C.:* Stable Organic Cation Salts: Ion Pair Equilibria and Use in Cationic Polymerization. Vol. 19, pp. 1–56.

Lee, C.-D. S. and *Daly, W. H.:* Mercaptan-Containing Polymers. Vol. 15, pp. 61–90.

Lipatov, Y. S.: Relaxation and Viscoelastic Properties of Heterogeneous Polymeric Compositions. Vol. 22, pp. 1–59.

Lipatov, Y. S.: The Iso-Free-Volume State and Glass Transitions in Amorphous Polymers: New Development of the Theory. Vol. 26, pp. 63–104.

Mano, E. B. and *Coutinho, F. M. B.:* Grafting on Polyamides. Vol. 19, pp. 97–116.

Mengoli, G.: Feasibility of Polymer Film Coating Through Electroinitiated Polymerization in Aqueous Medium. Vol. 33, pp. 1–31.

Meyerhoff, G.: Die viscosimetrische Molekulargewichtsbestimmung von Polymeren. Vol. 3, pp. 59–105.

Millich, F.: Rigid Rods and the Characterization of Polyisocyanides. Vol. 19, pp. 117–141.

Morawetz, H.: Specific Ion Binding by Polyelectrolytes. Vol. 1, pp. 1–34.

Morin, B. P., Breusova, I. P. and *Rogovin, Z. A.:* Structural and Chemical Modifications of Cellulose by Graft Copolymerization. Vol. 42, pp. 139–166.

Mulvaney, J. E., Oversberger, C. C. and *Schiller, A. M.:* Anionic Polymerization. Vol. 3, pp. 106–138.

Okubo, T. and *Ise, N.:* Synthetic Polyelectrolytes as Models of Nucleic Acids and Esterases. Vol. 25, pp. 135–181.

Osaki, K.: Viscoelastic Properties of Dilute Polymer Solutions. Vol. 12, pp. 1–64.

Oster, G. and *Nishijima, Y.:* Fluorescence Methods in Polymer Science. Vol. 3, pp. 313–331.

Overberger, C. G. and *Moore, J. A.:* Ladder Polymers. Vol. 7, pp. 113–150.

Patat, F., Killmann, E. und *Schiebener, C.:* Die Absorption von Makromolekülen aus Lösung. Vol. 3, pp. 332–393.

Penczek, S., Kubisa, P. and *Matyjaszewski, K.:* Cationic Ring-Opening Polymerization of Heterocyclic Monomers. Vol. 37, pp. 1–149.

Peticolas, W. L.: Inelastic Laser Light Scattering from Biological and Synthetic Polymers. Vol. 9, pp. 285–333.

Pino, P.: Optically Active Addition Polymers. Vol. 4, pp. 393–456.

Plate, N. A. and *Noah, O. V.:* A Theoretical Consideration of the Kinetics and Statistics of Reactions of Functional Groups of Macromolecules. Vol. 31, pp. 133–173.

Plesch, P. H.: The Propagation Rate-Constants in Cationic Polymerisations. Vol. 8, pp. 137–154.

Porod, G.: Anwendung und Ergebnisse der Röntgenkleinwinkelstreuung in festen Hochpolymeren. Vol. 2, pp. 363–400.

Pospíšil, J.: Transformations of Phenolic Antioxidants and the Role of Their Products in the Long-Term Properties of Polyolefins. Vol. 36, pp. 69–133.

Postelnek, W., Coleman, L. E., and *Lovelace, A. M.:* Fluorine-Containing Polymers. I. Fluorinated Vinyl Polymers with Functional Groups, Condensation Polymers, and Styrene Polymers. Vol. 1, pp. 75–113.

Rempp, P., Herz, J., and *Borchard, W.:* Model Networks. Vol. 26, pp. 107–137.

Rigbi, Z.: Reinforcement of Rubber by Carbon Black. Vol. 36, pp. 21–68.

Rogovin, Z. A. and *Gabrielyan, G. A.:* Chemical Modifications of Fibre Forming Polymers and Copolymers of Acrylonitrile. Vol. 25, pp. 97–134.

Roha, M.: Ionic Factors in Steric Control. Vol. 4, pp. 353–392.

Roha, M.: The Chemistry of Coordinate Polymerization of Dienes. Vol. 1, pp. 512–539.

Safford, G. J. and *Naumann, A. W.:* Low Frequency Motions in Polymers as Measured by Neutron Inelastic Scattering. Vol. 5, pp. 1–27.

Schuerch, C.: The Chemical Synthesis and Properties of Polysaccharides of Biomedical Interest. Vol. 10, pp. 173–194.

Schulz, R. C. und *Kaiser, E.:* Synthese und Eigenschaften von optisch aktiven Polymeren. Vol. 4, pp. 236–315.

Seanor, D. A.: Charge Transfer in Polymers. Vol. 4, pp. 317–352.

Seidl, J., Malinský, J., Dušek, K. und *Heitz, W.:* Makroporöse Styrol-Divinylbenzol-Copolymere und ihre Verwendung in der Chromatographie und zur Darstellung von Ionenaustauschern. Vol. 5, pp. 113–213.

Semjonow, V.: Schmelzviskositäten hochpolymerer Stoffe. Vol. 5, pp. 387–450.

Semlyen, J. A.: Ring-Chain Equilibria and the Conformations of Polymer Chains. Vol. 21, pp. 41–75.

Sharkey, W. H.: Polymerizations Through the Carbon-Sulphur Double Bond. Vol. 17, pp. 73–103.

Shimidzu, T.: Cooperative Actions in the Nucleophile-Containing Polymers. Vol. 23, pp. 55–102.

Shutov, F. A.: Foamed Polymers Based on Reactive Oligomers, Vol. 39, pp. 1–64.

Silvestri, G., Gambino, S., and *Filardo, G.:* Electrochemical Production of Initiators for Polymerization Processes. Vol. 38, pp. 27–54.

Slichter, W. P.: The Study of High Polymers by Nuclear Magnetic Resonance. Vol. 1, pp. 35–74.

Small, P. A.: Long-Chain Branching in Polymers. Vol. 18.

Smets, G.: Block and Graft Copolymers. Vol. 2, pp. 173–220.

Sohma, J. and *Sakaguchi, M.:* ESR Studies on Polymer Radicals Produced by Mechanical Destruction and Their Reactivity. Vol. 20, pp. 109–158.

Sotobayashi, H. und *Springer, J.:* Oligomere in verdünnten Lösungen. Vol. 6, pp. 473–548.

Sperati, C. A. and *Starkweather, Jr., H. W.:* Fluorine-Containing Polymers. II. Polytetrafluoroethylene. Vol. 2, pp. 465–495.

Sprung, M. M.: Recent Progress in Silicone Chemistry. I. Hydrolysis of Reactive Silane Intermediates. Vol. 2, pp. 442–464.

Stahl, E. and *Brüderle, V.:* Polymer Analysis by Thermofractography. Vol. 30, pp. 1–88.

Stannett, V. T., Koros, W. J., Paul, D. R., Lonsdale, H. K., and *Baker, R. W.:* Recent Advances in Membrane Science and Technology. Vol. 32, pp. 69–121.

Stille, J. K.: Diels-Alder Polymerization. Vol. 3, pp. 48–58.

Stolka, M. and *Pai, D.:* Polymers with Photoconductive Properties. Vol. 29, pp. 1–45.

Subramanian, R. V.: Electroinitiated Polymerization on Electrodes. Vol. 33, pp. 33–58.

Sumitomo, H. and *Okada, M.:* Ring-Opening Polymerization of Bicyclic Acetals, Oxalactone, and Oxalactam. Vol. 28, pp. 47–82.

Szegö, L.: Modified Polyethylene Terephthalate Fibers. Vol. 31, pp. 89–131.

Szwarc, M.: Termination of Anionic Polymerization. Vol. 2, pp. 275–306.

Szwarc, M.: The Kinetics and Mechanism of N-carboxy-α-amino-acid Anhydride (NCA) Polymerization to Poly-amino Acids. Vol. 4, pp. 1–65.

Szwarc, M.: Thermodynamics of Polymerization with Special Emphasis on Living Polymers. Vol. 4, pp. 457–495.

Takemoto, K. and *Inaki, Y.:* Synthetic Nucleic Acid Analogs. Preparation and Interactions. Vol. 41, pp. 1–51.

Tani, H.: Stereospecific Polymerization of Aldehydes and Epoxides. Vol. 11, pp. 57–110.

Tate, B. E.: Polymerization of Itaconic Acid and Derivatives. Vol. 5, pp. 214–232.

Tazuke, S.: Photosensitized Charge Transfer Polymerization. Vol. 6, pp. 321–346.

Teramoto, A. and *Fujita, H.:* Conformation-dependent Properties of Synthetic Polypeptides in the Helix-Coil Transition Region. Vol. 18, pp. 65–149.

Thomas, W. M.: Mechanism of Acrylonitrile Polymerization. Vol. 2, pp. 401–441.

Tobolsky, A. V. and *DuPré, D. B.:* Macromolecular Relaxation in the Damped Torsional Oscillator and Statistical Segment Models. Vol. 6, pp. 103–127.

Tosi, C. and *Ciampelli, F.:* Applications of Infrared Spectroscopy to Ethylene-Propylene Copolymers. Vol. 12, pp. 87–130.

Tosi, C.: Sequence Distribution in Copolymers: Numerical Tables. Vol. 5, pp. 451–462.

Tsuchida, E. and *Nishide, H.:* Polymer-Metal Complexes and Their Catalytic Activity. Vol. 24, pp. 1–87.

Tsuji, K.: ESR Study of Photodegradation of Polymers. Vol. 12, pp. 131–190.

Tsvetkov, V. and *Andreeva, L.:* Flow and Electric Birefringence in Rigid-Chain Polymer Solutions. Vol. 39, pp. 95–207.
Tuzar, Z., Kratochvíl, P., and *Bohdanecký, M.:* Dilute Solution Properties of Aliphatic Polyamides. Vol. 30, pp. 117–159.
Valvassori, A. and *Sartori, G.:* Present Status of the Multicomponent Copolymerization Theory. Vol. 5, pp. 28–58.
Voorn, M. J.: Phase Separation in Polymer Solutions. Vol. 1, pp. 192–233.
Werber, F. X.: Polymerization of Olefins on Supported Catalysts. Vol. 1, pp. 180–191.
Wichterle, O., Šebenda, J., and *Králiček, J.:* The Anionic Polymerization of Caprolactam. Vol. 2, pp. 578–595.
Wilkes, G. L.: The Measurement of Molecular Orientation in Polymeric Solids. Vol. 8, pp. 91–136.
Williams, G.: Molecular Aspects of Multiple Dielectric Relaxation Processes in Solid Polymers. Vol. 33, pp. 59–92.
Williams, J. G.: Applications of Linear Fracture Mechanics. Vol. 27, pp. 67–120.
Wöhrle, D.: Polymere aus Nitrilen. Vol. 10, pp. 35–107.
Wolf, B. A.: Zur Thermodynamik der enthalpisch und der entropisch bedingten Entmischung von Polymerlösungen. Vol. 10, pp. 109–171.
Woodward, A. E. and *Sauer, J. A.:* The Dynamic Mechanical Properties of High Polymers at Low Temperatures. Vol. 1, pp. 114–158.
Wunderlich, B. and *Baur, H.:* Heat Capacities of Linear High Polymers. Vol. 7, pp. 151–368.
Wunderlich, B.: Crystallization During Polymerization. Vol. 5, pp. 568–619.
Wrasidlo, W.: Thermal Analysis of Polymers. Vol. 13, pp. 1–99.
Yamashita, Y.: Random and Black Copolymers by Ring-Opening Polymerization. Vol. 28, pp. 1–46.
Yamazaki, N.: Electrolytically Initiated Polymerization. Vol. 6, pp. 377–400.
Yamazaki, N. and *Higashi, F.:* New Condensation Polymerizations by Means of Phosphorus Compounds. Vol. 38, pp. 1–25.
Yokoyama, Y. and *Hall H. K.:* Ring-Opening Polymerization of Atom-Bridged and Bond-Bridged Bicyclic Ethers, Acetals and Orthoesters. Vol. 42, pp. 107–138.
Yoshida, H. and *Hayashi, K.:* Initiation Process of Radiation-induced Ionic Polymerization as Studied by Electron Spin Resonance. Vol. 6, pp. 401–420.
Yuki, H. and *Hatada, K.:* Stereospecific Polymerization of Alpha-Substituted Acrylic Acid Esters. Vol. 31, pp. 1–45.
Zachmann, H. G.: Das Kristallisations- und Schmelzverhalten hochpolymerer Stoffe. Vol. 3, pp. 581–687.
Zambelli, A. and *Tosi, C.:* Stereochemistry of Propylene Polymerization. Vol. 15, pp. 31–60.

G. Govil, R. Hosur

Conformation of Biological Molecules

New Results from NMR

1981. 92 figures. Approx. 220 pages
(NMR, Volume 20)
ISBN 3-540-10769-X

Recent developments in NMR have made it an indispensable tool in biochemistry and molecular biology. With the advent of FT-NMR techniques, it has become possible to solve problems of sensitivity, resolution and assignments for medium size molecules, and to study their conformational structure and dynamics in solutions.
Availability of labelled compounds is contributing to a wider use of NMR for biological problems. Applications in studies of multimolecular systems, dynamics of cellular chemistry, biological control and regulation and short lived reaction intermediates at enzyme active sites have started to appear in literature and major contributions of NMR to the field of molecular biology can be expected in the future. (1247 references).
This article reviews recent trends and developments in conformational studies on biological systems using NMR. The first two chapters deal with the theoretical principles and NMR techniques used in conformational analysis. The final four chapters deal with applications to different classes of biological molecules: Nucleic acids and their components, amino acids, polypeptides and proteins, saccharides and polysaccharides and organisations in biomembranes. The emphasis is on basic principles and methodology in conformational analysis of biomolecules. Detailed coverage of the literature during the period from 1972 to 1980 is provided.

Springer-Verlag
Berlin
Heidelberg
NewYork

The International Journal for the Polymer Scientist covering all areas of Polymer Science

Polymer Bulletin

ISSN 0170-0839   Title No. 289

Editors:
Prof. H.-J. Cantow, Makromolekulare Chemie, Universität Freiburg. Prof. J.P. Kennedy, Dept. of Polymer Science, The University of Akron. Prof. T. Saegusa, Dept. Synthetic Chemistry, Kyoto University.

Editorial Board: H. Batzer, Basel; S. Cesca, San Donato Milanese; K. Dušek, Prague; P.J. Flory, Stanford, CA; J. Furukawa, Tokyo; J.E. McGrath, Blacksburg, VA; H.K. Hall, Jr., Tucson, AZ; M.L. Hallensleben, Hannover; H.H. Kausch, Lausanne; T. Kelen, Budapest; M. Kryszewski, Lódź; A. Ledwith, Liverpool; R.W. Lenz, Amherst, MA; E. Maréchal, Paris; J. Meißner, Zürich; A. Nakajima, Kyoto; G. and S. Henrici Olivé, Research Triangle Park, NC; N.A. Platé, Moscow; C.I. Simionescu, Bucureşti; S. Sivaram, Gujarat; D.H. Solomon, Melbourne; H. Tadokoro, Osaka; M. Takayanagi, Fukuoka; I. Uematsu, Tokyo; O. Vogl, Amherst, MA; C. Wippler, Strasbourg; H. Zahn, Aachen

Editorial Assistant: A. Heinrich, Springer-Verlag Heidelberg

Character: between the purely archival journals of full papers and "letter journals" consisting exclusively of short communications; length of papers, 4–8 pages

High-quality papers with an international spectrum: German-speaking countries, Eastern Europe and Japan 19% each; USA 13%; France 12%; other countries 18%

Competent referee system: rejection rate 35%

Rapid publication of papers: 3–6 weeks

50 reprints of each paper free of charge

No page charge

For subscription information and sample copy write to:
Springer-Verlag, Journal Promotion Dept., P.O. Box 105 280, D-6900 Heidelberg, FRG

G. V. Vinogradov, A. Y. Malkin

Rheology of Polymers

Viscoelasticity and Flow of Polymers

1980. 220 figures, 3 tables. XII, 467 pages
Moscow: Mir Publishers
ISBN 3-540-09778-3

This monograph presents a comprehensive review of theoretical and experimental investigations in the rheology of high-molecular-mass compounds and their solutions, the properties of which are examined in relation to their structure and composition. The book describes the most important features of the mechanical properties of polymers in the viscoelastic state. They are used as the basis for theoretically substantiated design calculation and selection of the processing equipment for molding various articles from polymeric materials. Much attention is devoted to a detailed consideration of results of investigations into the viscous properties of polymer melts and solutions, properties which govern polymer behaviour in the various manufacturing processes. The behaviour of fluid polymers under uniaxial extension conditions is treated in detail.

The book is intended for a wide range of scientists concerned with the physics, mechanics and physical chemistry of polymers, and also for technologists engaged in plastics processing, molding of articles from rubber mixes and production of fibers. The book will also be of assistance and guidance to students and postgraduates specializing in polymer science and technology.

A. Hebeish, J. T. Guthrie

The Chemistry and Technology of Cellulosic Copolymers

1981. 91 figures. XII, 351 pages
(Polymers, Volume 4)
ISBN 3-540-10164-0

Contents: The Homopolymeric Species. – Vinyl Graft Copolymerization onto Cellulose. – Radiation-Induced Grafting onto Cellulosics. – Grafting by Chemical Activation of Cellulose. – Grafting of other Types of Monomers onto Cellulose. – Grafting on Chemically Modified Celluloses. – Characterization and Properties of Cellulose Graft Copolymers. – Industrial Application of Cellulose Graft Copolymers.

The driving force behind the great scientific interest in copolymer science and technology is the search for products with useful, new or interesting properties. This monograph provides an informative account of new, improved cellulosic materials and the chemistry and technology involved in their production, as well as the first detailed description of grafted and modified celluloses.

Springer-Verlag
Berlin
Heidelberg
New York